WORLD ENERGY AND TRANSITION
TO SUSTAINABLE DEVELOPMENT

World Energy and Transition to Sustainable Development

by

Lev S. Belyaev

Oleg V. Marchenko

Sergei P. Filippov

Sergei V. Solomin

Tatyana B. Stepanova
and
Alexei L. Kokorin

Energy Systems Institute of the Siberian Branch of the Russian Academy of Sciences, Irkutsk, Russia

SPRINGER-SCIENCE+BUSINESS MEDIA, B.V.

A C.I.P. Catalogue record for this book is available from the Library of Congress.

ISBN 978-90-481-6137-9   ISBN 978-94-017-3705-0 (eBook)
DOI 10.1007/978-94-017-3705-0

Cover image: Block diagram of Lunar power system

Translated and updated version of:
L.S. Belyaev, O.V. Marchenko, S.P. Filippov, S.V. Solomin, T.B. Stepanova, A.L. Kokorin.
The world energy and transition to sustainable development. Novosibirsk, "Nauka", 2000,
269 p. (in Russian).
Original title: Мировая энергетика и переход к устойчивому развитию
Translators: V.P. Ermakova, M.V. Ozerova, A.S. Kiruta

Printed on acid-free paper

CONTENTS

Preface

The future attracts us by its uncertainty. It presents us with a variety of alternatives and generates a range of fears and hopes. Hardly anyone can remain indifferent to it. For scientists, it is as much a subject of cognition as are physical phenomena, chemical and biological processes, economic laws and many other spheres of intellectual curiosity. Investigations of the remote future are mainly of cognitive character. Forecasts for the near future must necessarily pursue more pragmatic — political, social, and economic —goals. The degree of depth and accuracy required of those forecasts depends on how concrete and specific the perceived problems are and how their solutions will effect the well-being of the planet and the people on it.

Interest in the remote future increased significantly in the 1960–70s, due in large part to studies initiated by "The Club of Rome" [1] which forcibly drew our attention to the limits of resources on our planet. The so-called energy crisis, caused by a sharp rise of oil prices in 1973 and 1979, was a stimulating factor for the energy sector.

Studies of long-term prospects for energy development in the world and its regions were begun in a number of national and international organisations. Particularly comprehensive studies were carried out at the International Institute for Applied Systems Analysis (IIASA) and summarised in the monograph [2]. Other works that are worthy of mention are [3–5]. In the 1980–90s an increasing number of countries and institutions took up similar studies. Recently, particular emphasis has been placed on energy problems in the relatively new context of mankind's transition to sustainable development.

The studies described in this book began in the late 1980s at the Siberian Energy Institute (SEI) of the Russian Academy of Sciences, Siberian Branch (known since 1998 as the Energy Systems Institute). They were from the beginning technology oriented. Their purpose was to comprehend long-term trends in the technological progress as a substantive basis on which to determine rational directions for development of the national energy sector.

A hypothesis was posed for possible formation in the 21st century of a technologically unified (multi-product) World Energy System (WES) as a way for development of centralised energy supplies to cities and industrial centres. It was conjectured that such a WES might be formed by creating very large (tens of gigawatts) energy centres that would produce a combination of electrical and thermal energy from nuclear energy and renewable energy sources (solar, tidal, etc) and also synthetic liquid and gaseous fuel from coal, shale and so on. Section 10.2 of this book is specifically devoted to describing the concept of a multi-product WES and analysing studies performed on a model of such a concept.

The Global Energy Model (GEM-10R) was devised to study the proposed World Energy System [8]. It divides the world into ten regions, allows for data about these regions over a time horizon that extends to the end of the 21st century, and specifies comparisons at time points 2025, 2050, 2075 and 2100. GEM-10 is a linear, optimisation, quasidynamic model with sufficiently detailed presentation of technologies for the whole cycle of energy production, conversion, transport and consumption. It incorporates environmental constraints (on emissions of CO_2, SO_2, NO_x and particulates), calculation of costs for regional energy development and operation, export-import dynamics of energy carriers, shadow prices, etc. After the model was provided with an extensive information on energy resources, demands, technologies and constraints, it became the main tool of our studies and proved to be a very efficient one.

In the majority of studies performed the world was divided into ten regions mentioned in the Table. This division principally coincides with the one normally accepted in global energy studies and differs only in greater territorial integrity of the regions (for example, regions of JK and AZ).

Table. Division of the world into regions

No.	Region	Symbol	Countries
1	North America	NA	USA (including Guam, Virgin Islands and Puerto Rico) and Canada.
2	Europe	EU	All countries of Europe (excluding republics of the former USSR) and Turkey.
3	Japan and Korea	JK	Japan and South Korea.
4	Australia and New Zealand	AZ	Commonwealth of Australia, New Zealand, New Caledonia.
5	The former USSR	SU	Russia and 14 republics (now countries) which were parts of the USSR.
6	Latin America	LA	Countries of South, Central America, Caribbean basin and Mexico.
7	Middle East and North Africa	ME	Afghanistan, Iran, all Asian countries to the west of Iran (excluding Turkey) and 6 countries of Africa (Egypt, Sudan, Libya, Tunisia, Algeria, Morocco).
8	Africa	AF	All countries of Africa excluding countries which are parts of region 7.
9	China	CH	China (including Taiwan and Hong Kong), Mongolia, North Korea, Laos, Vietnam, Cambodia.
10	South and Southeast Asia	SA	All countries of Asia excluding those in regions 3, 5, 7, 9 and also developing countries of Oceania.

In 1992–1999 the GEM-10R model was applied to perform the following extensive scope of studies on:

— the possible role and efficiency of the Earth's power supply from Space Power Systems including the Lunar Power System (LPS) [9–13];

— nuclear energy efficiency and possible scales of its development in the 21st century [12, 14, 15];

— the influence of global constraints on CO_2 emissions, on structure and economic indices of the world and regional energy [10, 12, 14];

— long-term prospects of energy development in the former USSR region and the Asian-Pacific region and their potential role in world energy [16];

— peculiarities of and requirements for long-term energy development which are caused by the necessity of transition of the world community to a new conception of sustainable development (under Russian Foundation of Basic Researches 96–82–18008) [14].

Furthermore, the authors of the book took part in the international project of IAEA "The overall comparative assessment of different energy sources (systems) and their potential role in long-term sustainable energy mixes" (1997–1999), in the project of European Commission EURIO-KIT (1997–1998) and other international projects.

A considerable body of information was accumulated during these studies and it was considered expedient to summarise it in a monograph. Based on the obtained results and experience of studies and also recently published works [17, 18, etc.] the authors pursued the following main goals in the book:

1. To analyse long-term tendencies and study possible alternatives of energy development in the world and its regions including:

— economically expedient scales of development of the existing and new energy technologies;

— distinctions in regional energy structures;

— possible scales of fuel exchange (trade) among the regions;

— distribution of global constraints on CO_2 emissions by region, their influence on energy structure and associated costs.

Such global and regional analysis became possible largely due to the unique peculiarities of the GEM-10R model.

2. To form notions of *energy for sustainable development* (by this term we will mean the entire concept of the energy sector meeting requirements and conditions for mankind's transition to sustainable development) and potentialities and conditions of its formation in different regions of the world. This problem is considered from the standpoint of searching for an optimal energy structure and scales of its development, providing, on the one hand, economic growth and improvement of the living standard of people, especially in developing countries, and on the other hand, reduction in the negative impact of man's activity on the environment up to a safe limit, which allows catastrophic consequences in the long-term future to be avoided.

3. To try to make *a forecast of "sustainable" energy development in the world and its regions* in the 21st century. For this purpose along with the "extreme" scenarios of energy development conditions (rigid constraints on CO_2 emissions, a moratorium on nuclear energy development, etc.) that were usually considered by the book authors and other researchers, two forecasted scenarios were generated. They are characterised by:

— a low energy consumption level, which at the same time turns out to be sufficient for economic growth and improvement of the standard of living in developing countries;

— financial and technological assistance of the developed to the developing countries;

— "mitigated" constraints on CO_2 emissions;

— introduction of "sound" constraints on nuclear energy development (including breeders) in the world regions;

— sufficiently justified constraints on introduction of new energy technologies, consumption of fossil fuel resources (primarily oil and natural gas), interregional fuel exchange and so on.

These forecasted or "credible" scenarios were generated by applying the experience and analysis of the results of previous studies and calculations on the GEM-10R. These scenarios are based in particular on the thesis which the authors arrived at: "In the 21st century there will be no prevailing energy resource (with a fraction of more than 30 or 35%), as was the case before". It may be expected that all main energy resources (coal, oil, natural gas, nuclear energy and renewable energy sources) will be used in certain proportions.

Despite the conventional character of such forecasts, they can give a realistic view of the future energy, its possible influence on the environment, needed economic expenditures, etc. Subsequently they will surely require adjustment in striving for perfection.

The following methodological "innovations", improving reliability of the results are applied in the described studies:

— the original GEM-10R model, optimising the world energy as a whole, is used;

— demands for final energy forms (electrical, thermal, mechanical and chemical ones) are considered, allowing the simulation of the whole energy chain from primary energy production to final energy consumption and competition (substitution) of energy carriers;

— a modified method of direct calculations is proposed to determine minimum necessary energy demands for the long-term future;

— division of reserves of primary energy resources (renewable and non-renewable) into 7–9 cost categories (including very expensive ones) is foreseen, making it possible to consider the nonlinear dependence of available resources on their cost.

Chapter 1 presents a survey of the current state and trends of the world energy development as an object of the described studies. Statistical data on primary energy production, energy consumption, economic growth rates, etc. are given and analysed. The chapter also gives a brief description of previous forecasts of world energy development for the medium- and long-term future.

Chapter 2 describes an applied methodology of world energy study. Requirements for energy resulting from the principles of sustainable development are analysed; a general pattern and peculiarities of studies are considered. Mathematical description of the GEM-10R model is given.

Chapters 3–5 consider external conditions of world energy development: forecasts of energy consumption in the world regions, estimates of energy resources, characteristics of energy technologies. These data were used to generate the studied scenarios in Chapter 6.

Analysis of calculation results for different scenarios is carried out in Chapters 7 and 8. Consideration is given to trends in the use of primary energy resources in the world and its regions, rates of cheap fossil fuel depletion and possible changes in its prices, scales of synthetic fuel production, peculiarities of energy development in the regions and interregional exchange of energy resources.

Chapter 9 is devoted to the additional analysis of results in terms of sustainable development requirements. Primary attention is paid to the problem of greenhouse gas emission reduction. Supplementary costs to mitigate emissions are compared with GDP of the regions and with predicted damage due to change in the global climate. Influence of these costs on macroeconomic indices of different regions is estimated. Diverse options of supplementary cost sharing among the regions and the arising problems are considered.

Chapter 10 deals with the mix of new energy technologies and efficiency of their use in the 21st century. Conditions and necessary terms of their implementation are analysed, technologies are ranked in efficiency, priorities in their development are determined. A special section is devoted to the mentioned technologically unified WES.

In conclusion the main results obtained and some objectives of further studies are briefly formulated.

Participation of the authors in writing the book is as follows:

L.S. Belyaev — Preface, Chapters 2, 6, 7, 8 and 10, Sections 3.1 and 5.5, Conclusion;

O.V. Marchenko — Preface, Chapters 1, 7, 8 and 9, Sections 2.1 and 3.4;

S.P. Filippov — Chapters 2, 4, 5, 7, 8 and 10, Section 1.1;

S.V. Solomin — Chapters 7, 8 and 10, Sections 4.1, 4.4 and 5.2;

T.B. Stepanova — Section 3.2;

A.L. Kokorin — Section 3.3.

The authors are grateful to Corresponding Member of RAS S.N. Vassilyev, Prof. Yu.D. Kononov and Prof. B.G. Saneev for their manuscript review and valuable comments which promoted its improvement. It is also their pleasant duty to express thanks to V.P.Ermakova, M.V.Ozerova and A.S.Kiruta for translating the text from Russian into English and also to E.G. Lapteva, O.M. Kovetskaya, G.G. Bonner, L.K. Rogova and L.M. Shiryaev for great work on preparing the manuscript for publication.

The authors would like to thank Mr. Edwin Beschler for his assistance in language editing, and Mrs. Betty van Herk, of Kluwer Academic Publishers, for her assistance in the publication of this volume.

Part I.

METHODOLOGY OF STUDIES AND EXTERNAL CONDITIONS OF ENERGY DEVELOPMENT IN THE 21st CENTURY

Chapter 1

WORLD ENERGY: STATE OF THE ART AND TRENDS IN DEVELOPMENT

Long-term prospects for world energy development are based on the analysis of quantitative indices that describe the state of the art and provide a "benchmark" for our predictions, and also on the examination of past development trends. Towards this end, Chapter 1 presents a short survey of principal statistical indicators of world energy development that are important for the present study.

We consider indices for the world as a whole, individual countries, ten regions, described in the GEM-10R model, and three more large aggregations — developed countries (North America, Western Europe, Japan, South Korea, Australia, and New Zealand), countries with transition economies (the former Soviet Union and East European countries) and developing countries (rest of the world). Statistical data, covering the time span to the middle or end of the 1990s, are applied. Indices for the year 1990 are taken as basic for two reasons: first, it was the time of GEM-10R construction and performance of the first series of computations on it; second, the early 1990s saw drastic changes in indices of the economy and energy development in the countries of the former USSR and Eastern Europe. These changes should naturally be taken into consideration in long-term forecasting. However, the economic development level of these countries, energy consumption structure, energy intensity of product output, etc. are best characterised by the pre-crisis values of indices.

1.1. Definitions

The energy sector consists of a combination of processes of energy conversion and transportation from sources of natural energy resources to consumers of energy services. The possibility, and necessity, to treat it as a united system are due to first, close technological links of enterprises, implementing successive stages of energy conversion, and second, substitution of different energy forms and energy carriers.

As to conversion stages, energy is divided into primary, secondary and final forms (Table 1.1).

According to the accepted classification in GEM-10R *primary energy* is the energy extracted directly from the environment, i.e. it is a thermal equivalent of produced coal, crude oil, natural gas, collected biomass, kinetic energy of water or wind, energy of solar radiation at the collector plane, thermal energy of hot water or steam in the Earth's depth.

The primary energy produced is converted to *secondary energy* (secondary energy carriers) in the form of electrical energy or fuel (gasoline, fuel oil, methanol, hydrogen, substitute natural gas, synthetic motor fuel, etc.).

Table 1.1. Energy forms and stages of its conversion

Process (*energy*)		Examples
Production (mining)	Oil production	Coal mining
Primary	*Crude oil*	*Coal*
Primary energy conversion	Oil refinery	Power plant
Secondary	*Gasoline*	*Electrical energy*
Final energy production	Internal combustion engine	–
Final	*Mechanical energy*	*Electrical energy*

Electrical final energy is a portion of all electrical energy produced and transported to consumers that can not be replaced by other energy forms (electrical energy for lighting, communication systems, computers, etc.).

Thermal final energy is the heat produced and transported to consumers by heating systems or consumed in industrial processes.

Mechanical final energy is equal to kinetic energy of motors used in transport.

Chemical final energy is equal to the thermal equivalent of oil, oil products and natural gas used as feedstock by the chemical industry.

Use of final energy to move an automobile (mechanical energy) or to power a lighting system renders such *energy services* to consumers as mobility or intensity of illumination.

Note that other definitions which differ from the indicated ones are often applied in publications. This fact must be kept in mind when data from different sources are compared. The primary energy of renewable energy sources (sun, wind, geothermal energy) sometimes is equated with either electrical or thermal energy produced from them or primary energy consumed by some conditional fossil-fired power plant that produces the same quantity of electrical energy (fuel equivalent of primary energy).

If the efficiency of this power plant is taken equal to 33%, a unit of primary energy consumed by a hydropower plant with an efficiency of 85% is equivalent to 0.85/0.33=2.58 units of primary energy of fuel. For a solar power plant with a 10% efficiency this relation is 0.10/0.33=0.30.

Final energy is often understood as electrical energy and fuel, supplied to consumers, and the mechanical work and thermal energy produced from them are referred to as *useful energy*.

When the whole energy sector (from production of primary energy resources to energy consumption and provision of energy services) is studied (modelled), these distinctions in terminology are inessential. However, due to the variety and diversity of consumption processes the energy production sector is considered, as a rule, separately. In this case it is important to agree on what is meant by final energy. The selected four types of final energy (electrical, thermal, mechanical and chemical) set a boundary between the energy production and the consumption sectors. They are well suited for modelling energy development, since they allow (as distinct from the case when the final energy defined is electrical energy and fuel) the modelling (optimisation) of substitution of different energy carriers for final consumers (e.g. coal, gas, biomass and electrical energy for heat production).

1.2. Statistical indicators

In the year 2000 the world primary energy consumption was about 15 billion tons of coal equivalent (tce) per year (2.5 tce per capita). Since the early 20th century energy consumption has risen by a factor of 11, whereas the population has increased 3.7 times [18–22]. The maximum rate of energy consumption growth (4–5% per year) which coincided with the "population explosion" in 1950–1970 has declined to 1–3% per year in the ensuing 25–30 years (Figure 1.1, Table 1.2). During these years the world average energy consumption per capita has stabilised at a level of 2.2–2.4 tce.

Table 1.2. Growth rates of the world's population, total and specific (per capita) energy consumption, % per year

Year	Population	Energy consumption	
		Total	Specific
1900–1930	0.7	2.1	1.4
1930–1950	1.0	1.3	0.3
1950–1970	1.9	4.7	2.7
1970–1990	1.9	2.0	0.1
1990–1996	1.5	1.5	0.0

In the pre-industrial era (until the middle of the 18th century) energy demands of people were met basically by energy resources and energy flows directly available in nature (wood, wind, water) and also by the muscular force of men and animals. The average energy consumption per capita did not exceed 0.4 tce per year [18]. The succeeding time span was characterised by two essential structural changes in the world energy.

The first was caused by invention and extensive use of a steam engine burning coal as fuel. The second was conditioned by appearance of an internal combustion engine, giving rise to a sharp increase in the demand for liquid fuel and correspondingly oil production.

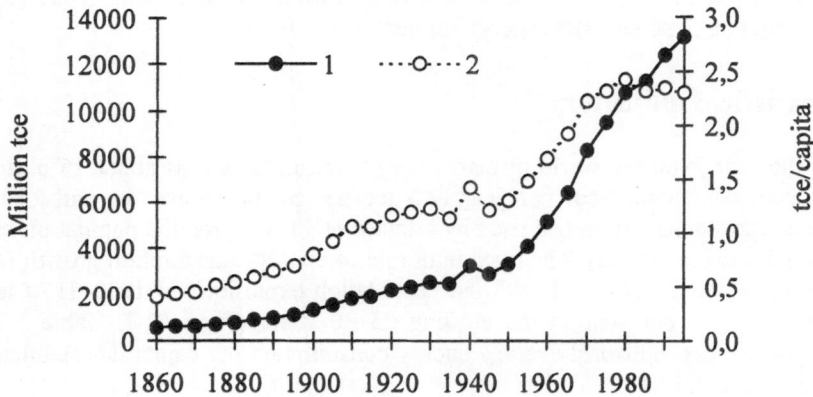

Figure 1.1. The world total (1) and specific (2) annual energy consumption.

The history of the world energy development for the last 1.5 centuries demonstrates successive change of the dominating energy carriers: wood – coal – oil (Figure 1.2). Many investigators have made multiple attempts to explain, quantitatively describe and predict regularities in the future change of energy structure.

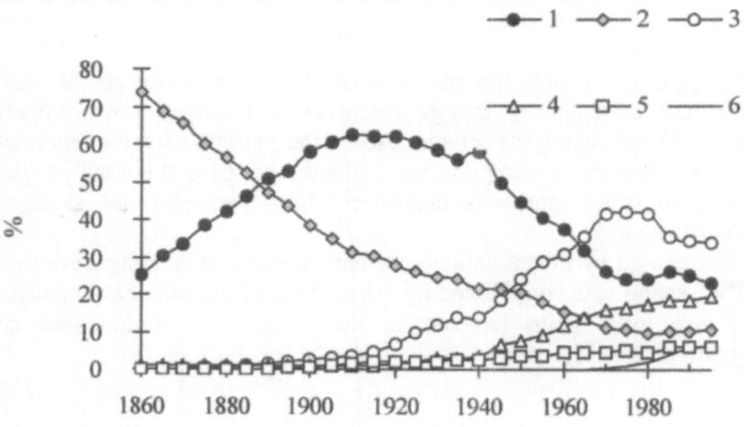

Figure 1.2. World energy consumption structure.
1 – coal, 2 – biomass, 3 – oil, 4 – gas, 5 – hydropower (fuel equivalent), 6 – nuclear energy.

As evidenced by the analysis of factual data of several hundreds of examples from different branches of engineering, economy and social life, in the course of development each technology (or in a broader sense, innovation) is characterised by the following sequential stages [23]: 1) invention; 2) demonstration; 3) penetration of the market; 4) wide use; 5) saturation; 6) replacement by more advanced technology. Therefore, the share of the new technology on the market (the share of product or service provided by this technology) in terms of time is represented by the S-shaped curve.

J.C. Fischer and R.H. Pry [24] proposed a simple mathematical model, describing substitution of one technology by another. In the model the growth rate of the fraction F of a new technology is assumed proportional to the product of fractions of the considered and the competing old technologies, i.e., $F = \text{const } F (1 - F)$. In this case dependence $F(t)$ is the S-shaped curve (a logistic function) and the value $\ln (F/(1 - F))$ is linearly dependent on time t. If the curve parameters on its initial section are known, one can forecast terms and scales for the new technology to be introduced.

Extension of the substitution model to the case of n technologies [25] made it possible to quantitatively describe and forecast the technical development process as an alternating sequence of technologies which appear, penetrate the market, reach saturation and are displaced from the market.

C. Marchetti [26] advanced a hypothesis according to which the primary energy consumption structure and the sequence of change of a dominating energy carrier were governed by the same regularity. Processing of statistical data showed the substitution model to embody great prognosticating capabilities. For example, the coefficients, which were determined from information for the 20–year period, 1900–1920, were applied to describe the primary energy consumption structure for a much longer time horizon (1860–1974) with an extremely high accuracy [2].

Numerous investigations were carried out on the model to forecast a continuing process of change in the dominating energy carriers (wood – coal – oil) in the form: gas – nuclear energy – solar energy with transition to preferential use of hydrogen as a secondary energy carrier [27–29].

In the late 1970s the substitution model predicted practically complete displacement of biomass from the world energy balance, decline of the coal fraction to 10%, increase of the gas fraction to 50%, and increase of the nuclear energy fraction to 10% by the year 2000. However, the simultaneous studies on long-term prospects for the world energy development which were performed on a system of mathematical models showed [2] that in the following 20-year time span the fraction of coal and natural gas was expected to stabilise, i.e., the earlier revealed regularities of a sequential change in dominating energy carriers would be violated in the future. This conclusion was confirmed by the factual data (Figure 1.2).

Consideration of regional specific features is of great importance for the analysis of indices and trends of the world energy development. This is conditioned by unevenness in allocation of natural resources over the territory and to a greater extent, essential distinction in the economic development levels of the world's countries.

The gross domestic product (GDP) that represents the total cost (in market prices) of end-use goods and services produced in a territory is the most general aggregated index, describing the economy of a country or region. GDP per capita is a quantitative estimation of the economic development level and with a certain degree of conventionality, the average living standard of the population.

GDP of different countries can be compared in terms of the common monetary unit (for example, US$). However, with the use of market exchange rates the results may appear to be poorly comparable (in particular, between developed and developing countries). This is due to the restricted convertibility of some currencies, the sizeable difference in domestic and world prices of individual goods, and national distinctions in the ways the economic activity indices are taken into account. A methodology of purchasing power parities for use in the international comparison of GDP was elaborated to overcome these difficulties [30]. According to this methodology the national GDP is divided into some groups of homogeneous goods, and relationships between the national currency and the one chosen as basic (the rate with regard to purchasing power parity) are calculated by the average prices of goods in different countries. Then the national indices of the groups of goods are converted to "internationally comparable" indices and aggregated once again to GDP. For example, in 1990 GDP of the former USSR in market prices made up US$ 2.8 thousand per capita [31, 32]. If the purchasing power parity was taken into consideration, according to different estimations [17, 33, 34] GDP increased from US$ 6.4 to 7.4 thousand per capita.

Comparison of GDPs is very tedious work and carried out within the framework of international projects implemented with an interval of 3–5 years for a limited number of countries. Hence, in parallel with GDP, considering purchasing power parity, use is made of more accessible data on GDP in market prices. This is well justified in some cases (for example, in comparing only countries with a developed market economy, when the national exchange rate with purchasing power parity slightly deviates from the market exchange rate, as well as in studying GDP and its derivatives versus time for one and the same country).

For more convenient comparison of results, GDP values with purchasing power parity are applied in constant 1990 US$. The cost indices from the information sources used were converted to these units in advance.

The considered world regions are characterised by both essential distinctions in economic development levels (per capita GDP in the North America region is 16 times higher than the corresponding index for the least developed Africa region) and even greater distinctions in energy production and consumption indices (a 30-fold gap in the per capita energy consumption between the same regions). On the whole the developing countries with a population of above 3/4 of the world produce as little as 1/3 of the gross world product and consume 1/4 of its energy [17, 30, 31, 33, 35], (Table 1.3). Regional distinctions have a far lesser effect on the other specific index – the energy GDP ratio. In the regions (EU, JK, AZ, LA, ME, AF, CH, SA), combining both developed and developing countries the energy GDP ratio (in terms of primary energy) lies in a relatively narrow interval of 0.24–0.47 tce/1000 US$ of GDP. Higher values of the ratio are typical of the NA region and particularly, the former USSR. This is explained by excessive energy consumption both for lack of incentives to energy conservation and in view of geographical and natural–climatic features.

Table 1.3. Characteristics of the world regions (as of 1990)

Region*	Popula-tion,	GDP		Primary energy, tce/capita		Energy-GDP ratio, tce/thousand US$
	Million	trillion US$	thousand US$/capita	Production	Consump-tion	
NA	276	5.80	21.0	11.0	12.4	0.59
EU	554	6.46	11.7	3.1	5.0	0.43
JK	170	2.48	14.6	0.9	4.6	0.32
AZ	21	0.36	17.1	11.6	7.6	0.45
SU	289	1.85	6.4	8.6	7.2	1.13
LA	448	2.11	4.7	2.0	1.5	0.32
ME	271	1.14	4.2	6.8	2.0	0.47
AF	502	0.65	1.3	0.8	0.4	0.31
CH	1284	2.69	2.1	0.9	0.9	0.43
SA	1477	2.38	1.6	0.4	0.4	0.24
Developed	860	14.12	16.4	5.6	7.8	0.47
FSU & EE	404	2.53	6.3	7.0	6.4	1.02
Developing	4028	9.27	2.3	1.2	0.8	0.34
World	5292	25.9	4.9	2.4	2.3	0.48

* Division of the world into regions – see Preface; FSU & EE – former Soviet Union and countries of Eastern Europe.

Whereas total energy consumption is determined primarily by the economic development level, the structure of energy consumption, and in particular production, depends largely on available energy resources in the territory. The main energy producers (and consumers) are represented by the regions: NA, SU and EU (60% of the world coal production, 40% of oil and 79% of natural gas); the ME region concentrates 33% of the world oil production; the CH region – 25% of coal ([31], Figure 1.3).

World coal export in 1995 made up 13% of its production volume. The main exporters are Australia, the USA and the countries of Africa. The importers are Japan and Western Europe. Natural gas is also consumed mainly in the producing countries (import totals 16% of production). Oil and oil products rank first in the world trade in energy resources (56 % of world oil production) [20]. First of all oil is exported from the Middle East countries to Western Europe, the USA and Japan.

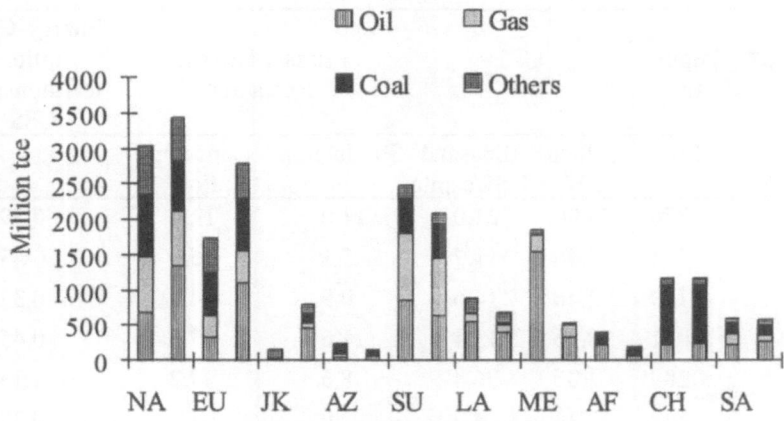

Figure 1.3. Production (left-hand column) and consumption of primary energy (1990).

Developed countries (considered as one region) import 28% of primary energy consumed by them, including 63% of their oil consumption.

Fossil fuel (coal, oil, gas) provides more than 80% of world primary energy. Nuclear energy (6%) is used mainly in the developed countries, biomass — in the developing countries (practically entirely for production of thermal energy — firewood for cooking and heating). The share of hydropower in the world energy balance makes up 2.5% (6% in fuel equivalent), the fraction of the other renewable energy sources (wind, sun, geothermal energy) is less than 1% ([36, 37], Figure 1.4).

Primary and secondary energies (mainly liquid fuels, obtained from oil refining) are converted to final energy with the efficiency of 40–50% ([37], Figure 1.5). The average efficiency values for electrical energy production amount to 30–40%, for thermal — to 60–70%, mechanical — to 20–25%. In all the world regions energy resources consumed for electrical energy production vary in the interval of 8–13%. Developed countries are characterised by a higher share of mechanical and chemical energy, developing countries by that of thermal energy.

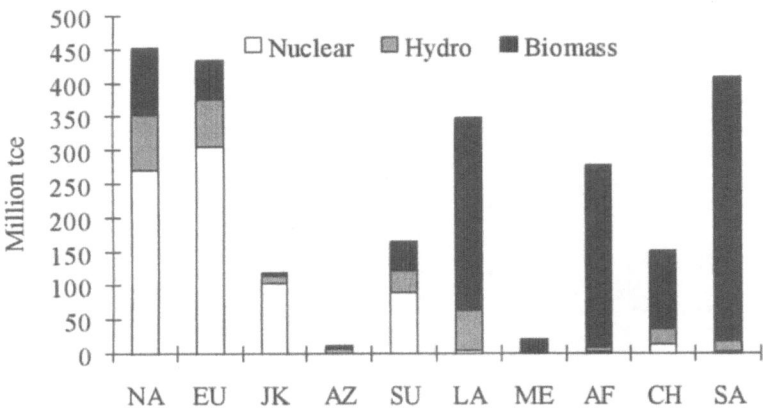

Figure 1.4. Consumption of nuclear and renewable energy (1990).

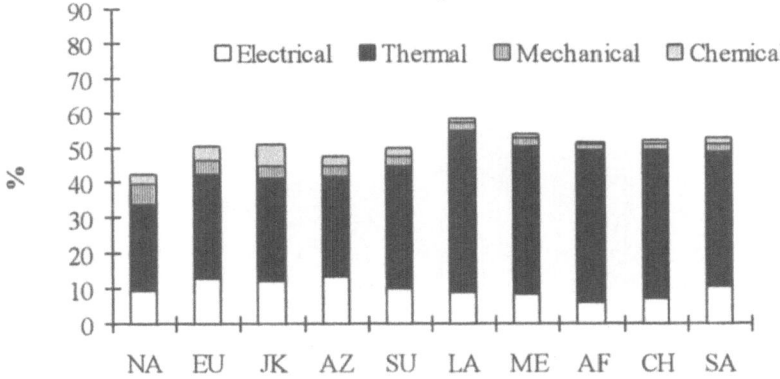

Figure 1.5. Structure of final energy production (1990), % of primary and secondary energy consumed.

Most countries use coal, nuclear energy and hydropower for electrical energy production; the fraction of liquid and gaseous fuel is high in the JK region as well as in the oil exporting regions — ME, SU and LA ([37], Figure 1.6). Developed countries use mainly liquid and gaseous fuel for thermal energy production, whereas developing countries use mainly biomass. In the most economically underdeveloped region of AF the biomass fraction (firewood) makes up about 80% (Figure 1.7) in the thermal energy production. The larger part of mechanical and chemical energy is produced using oil and oil products.

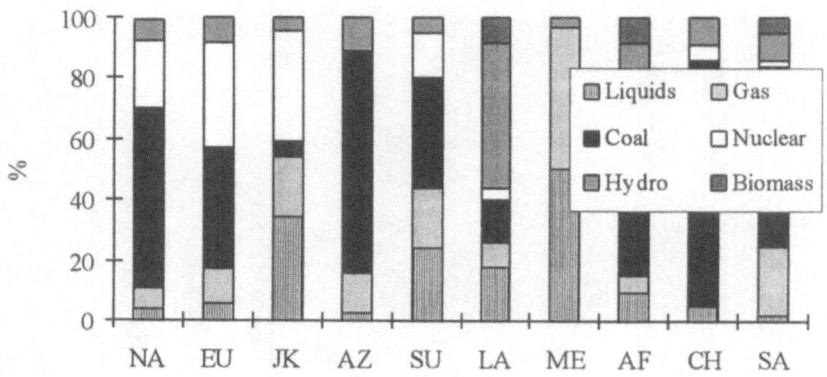

Figure 1.6. Structure of electrical energy production (1990).

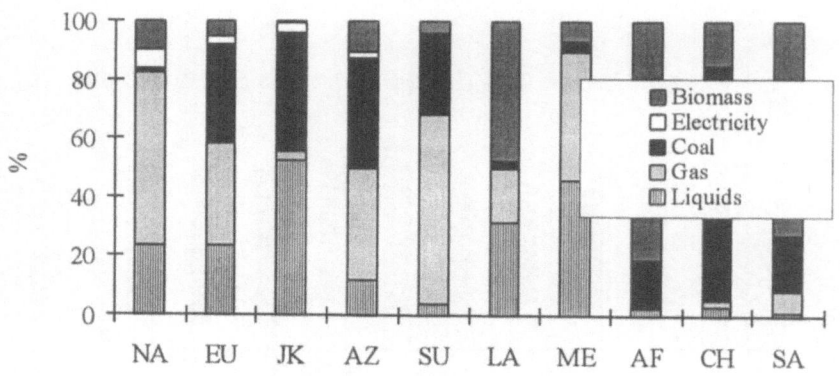

Figure 1.7. Structure of thermal energy production (1990).

The main consumers of energy resources are industry and the residential sector (private houses and commercial buildings). The share of agriculture does not exceed 5%; and in the developed countries the share of transport that consumes mainly liquid fuel is high (24%). In the developed countries the residential sector consumes 57% of electrical energy, in the developing — 38% ([38–40], Figures 1.8 and 1.9).

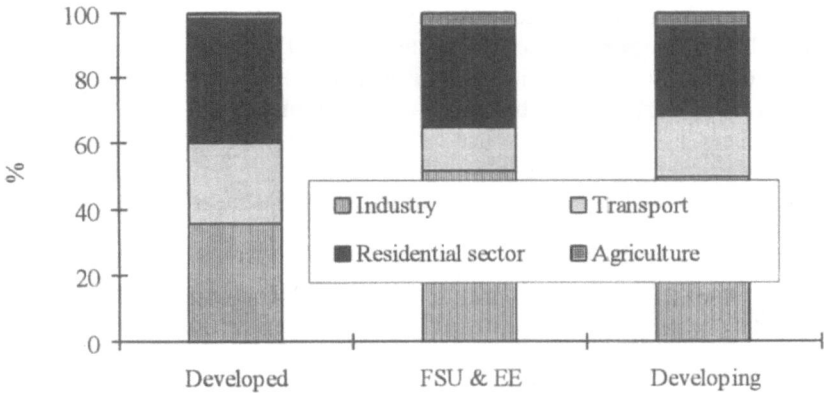

Figure 1.8. Structure of primary energy consumption by sector of the economy (1990).

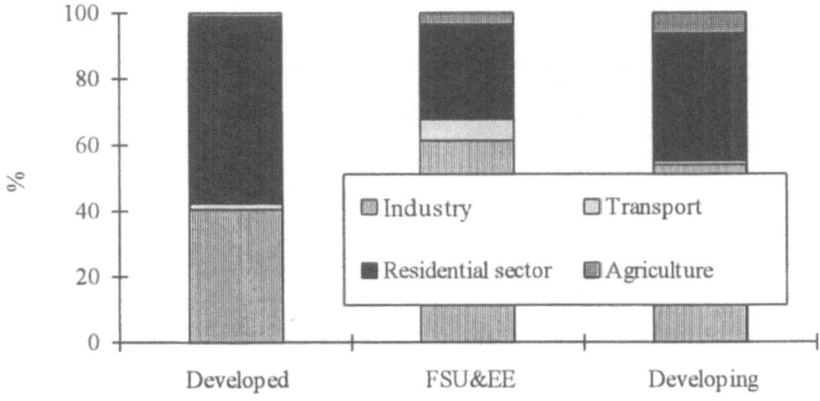

Figure 1.9. Structure of electrical energy consumption by sector of the economy (1990).

Production, transport, processing and combustion of fossil fuel cause energy flow discharges and emissions of different substances — oxides of carbon, sulphur, nitrogen, particulate — into the environment. Their negative impacts on nature and health of people is of local or regional character (air and water pollution near energy plants, local changes in moisture and climate, acid rains). At the same time energy may cause global changes in the climate by increasing concentrations of the so-called greenhouse gases, particularly CO_2.

In 1990 world carbon emission in the form of CO_2 caused by fuel processing and burning made up 5.8 billion t, i.e. 0.47 t C per 1 tce of primary energy consumed. This index was within the range from 0.39 (Latin America) to 0. 63 t C per 1 tce (Africa). In recent years the total world CO_2 emission has changed insignificantly despite growing energy consumption ([31], Figure 1.10).

This is caused by "decarbonisation" (decrease in emissions per unit of energy consumed) of the world energy balance, which has manifested itself as a global trend in world energy development since 1850 [23]. The reason for this phenomenon is change in the energy structure in favour of fuel with lower carbon and higher hydrogen content as well as in favour of hydropower and nuclear energy (see Figure 1.2).

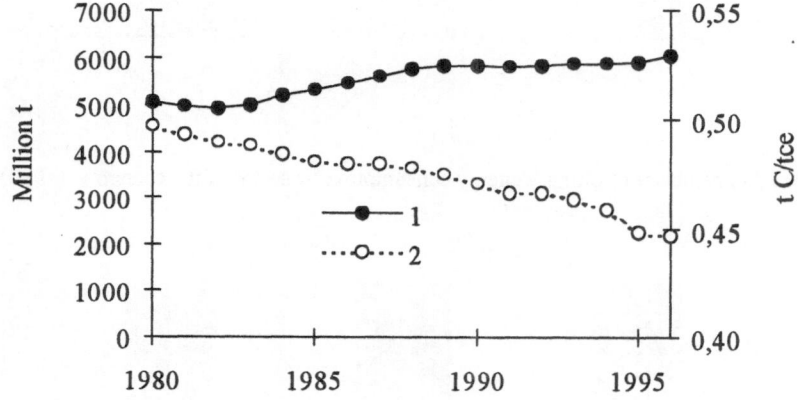

Figure 1.10. World total (1) and specific (2) (per unit of primary energy) CO_2 emission.

1.3. Energy and economy

World energy development in the long term will be determined by both overall economic growth and technological progress. Unevenness and cyclic character of development associated with technological progress may be considered in terms of changing technological structures (TSs), which represent an aggregate of technological processes developing synchronously and maintaining their integrity [18, 41].

In the 19–20th centuries, five technological structures appeared and developed. In the 1990s, developed countries entered the fifth technological structure, the core of which consisted of electronic industry, computers, software, information services and telecommunications. Based on the theory of technological structures the regularities of long-term changes in the world energy were revealed: per capita energy consumption increases almost twice when passing to the next structure; energy price rises; the stages of a new TS development are attained by some growth in energy-GDP ratio; replacement of the old TS by a new one is attained by fast decline in energy intensity [18].

The theory of technological structures allows one to explain and, to some extent, to forecast long-term changes in energy consumption of developed countries. Many developing countries with a low level of economic development are characterised by a great gap between the living standard of the rich and poor as well as a heterogeneous mix of principally different types of production and consumption — from primitive pre-industrial to the latest. For instance, Brazil, by a descriptive expression [5], "consists of one country within the other (Belgium inside India)". Therefore it is sensible to use simpler descriptions in long-term forecasts which pay great attention to the developing countries.

In a schematic form the process of economic development can be considered as a consequent passing of the countries through the following stages: 1) pre-industrial; 2) industrialisation; 3) industrially developed society; 4) post-industrial. Each of these stages corresponds to definite regularities of change in the population number, economic indices and energy consumption ([42], Table 1.4).

Table 1.4. Classification of development stages

Development stage	Industrial structure	Growth rate		
		Population	Economy	Energy consumption
Pre-industrial	Handicraft workshops, light industry	Low	Low	Low; non-commercial energy
Industrialisation	Metallurgy, chemistry	Medium	High	High
Industrially developed society	Car industry, home electronics	Low	Medium	Medium; oil, electricity
Post-industrial	Electronics and informatics	Negative	Low	Low or negative; energy diversification

Let us consider the economic growth rates for a 40-year period (1950–1990) in the example of seven developed countries (the USA, Canada, Great Britain, France, Germany, Italy, Japan), the former USSR and four developing countries (China, India, Brazil, Egypt); about 60% of world population lives in these countries. Since annual GDP growths vary greatly, the data [34] were preliminarily smoothed out (sliding average GDP values by a 5-year period).

Figures 1.11 and 1.12 present the relation between GDP growth rate and the economic development level (GDP per capita) for two groups of countries. It is seen that economic development rates undergo sharp changes and differ both for different countries and for different development periods of one and the same country. The detailed analysis [43] shows that the economic growth rates are affected by a number of factors, starting with weather conditions and finishing with medical care.

It is rather hard to consider them in long-term studies. For the purposes of the work it is sufficient to consider the most common quantitative (or even qualitative) regularities.

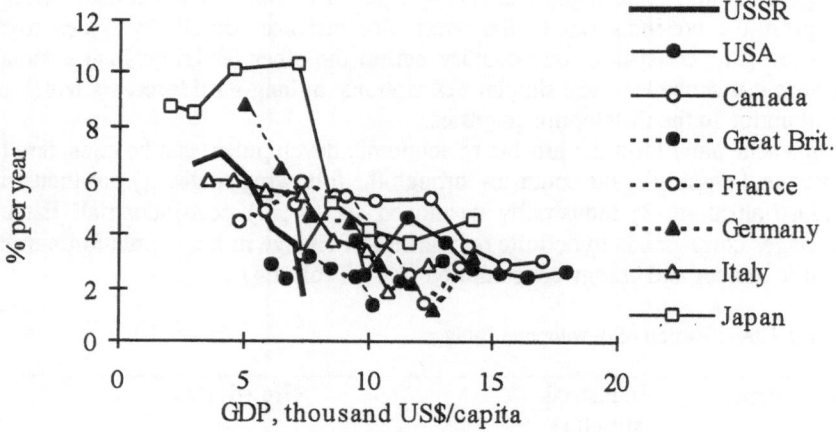

Figure 1.11. Relation between the GDP growth rate in developed countries and the USSR and GDP per capita.

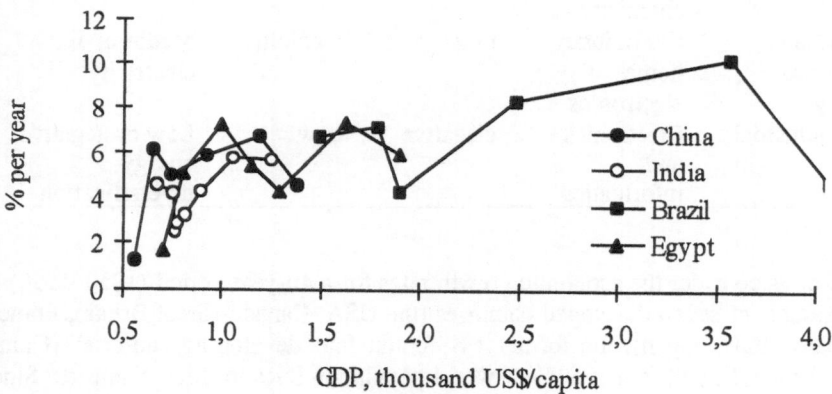

Figure 1.12. Relation between the GDP growth rate in developing countries and GDP per capita.

It can be noted that despite a wide scattering the curves (Figures 1.11 and 1.12) follow the general regularities, reflected in Table 1.4:

— GDP growth rate in the industrial countries and those passing to the post-industrial stage decreases to a value of about 2% per year;

— developing countries pass through the stage of accelerated economic development (6–10% per year), with subsequent decelerated growth.

It is interesting to consider dynamics of economic development of developing countries in terms of their "lagging" behind the more developed countries. Denote by ΔT the time span during which country 1 with GDP per capita $g_1(t)$ at the time instant t will reach this index of a more developed country 2 with specific GDP $g_2(t)$. Then,

$$\Delta T = \ln[g_2(t)/g_1(t)] / \ln[(1+\beta)/(1+\lambda)] \approx [1/(\beta-\lambda)]\ln[g_2(t)/g_1(t)],$$

where β and λ are the growth rates of GDP and population of country 1 at the time instant t.

Since the rates of economic growth vary greatly, as was pointed out above, it is expedient to replace β with an averaged value (average potential growth rate). In accordance with the trend of decreasing economic growth rate with rising GDP per capita g, select function $\beta(g)$ in the form of a linear relationship so that the potential growth rate decreases from 6% per year at $g=0.5$ thousand US$/capita to 2% per year at $g=18$ thousand US$/capita.

The calculations show that by 1990 the gap between the six considered developed countries and the USA was from 4 (Canada) to 12 years (Italy), 25 years for the USSR, for the developing countries from 45 (Brazil) to 72 (India) years (Figure 1.13). During 1995–1990 the average rate of the USA population growth equalled 1.2% per year, that of GDP — 3.1 and specific GDP — 1.8% per year. For India for the same period of time the growth rates made up: 2.1% per year for the population, 4.1% for GDP and 1.9% per year for specific GDP. Due to closeness of the specific GDP growth rates the gap between the USA and India varied within 60–70 years practically without reduction.

Brazil with an average growth rate of specific GDP of 3% yearly (the population — 2.6, GDP — 5.7% yearly) managed to reduce the gap with the USA from 72 to 45 years. However after reaching the value of US$ 4 thousand /capita by 1980 the specific GDP stabilised due to a decrease in GDP growth rate to 2% per year. This led to increasing the gap between Brazil and the USA.

Data analysis for a longer time span [43] shows that despite the uneven rates of economic growth the order of countries ranked by the specific GDP values in most cases changes within the group of countries with similar levels of economic development only. Thus, for instance, the specific GDP of Germany in 1913 made up 72% of the USA level, in 1992 — 90%, that of France — 95 and 73% respectively. At the same time the specific GDP of India both in the beginning and in the end of the 20th century was many times less (12% in 1913 and 6% in 1992).

In the first half of the 20th century the specific GDP growth for successfully developing countries made up 1.5 % per year, in the second half — 3.5% per year with a subsequent drop to 1–2% per year. After 1950 only a limited number of countries (Japan, then South Korea, Taiwan, Singapore, Malaysia, Thailand) demonstrated very high rates of specific GDP growth (5–8% per year), reached developed European countries or approached them and entered the advanced stage of economic growth.

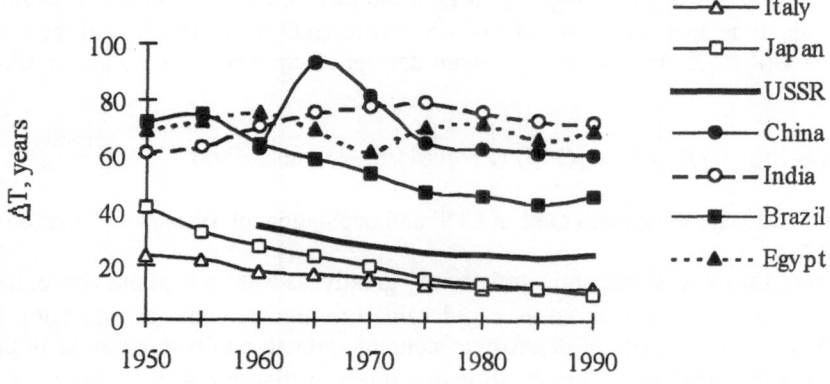

Figure 1.13. The gap between some countries and the USA by specific GDP.

In 1950–1990 the specific GDP of Japan increased ten-fold at its average growth rate of 5.9% per year, whereas Great Britain increased only 2.4-fold at the growth rate of 2.2% per year. Partially this distinction can be explained by a low initial level of specific GDP of Japan in 1950. However, as seen from Figure 1.11, Japan's economy growth rates surpassed the corresponding growth rates in Great Britain at the same development level. This testifies to the fact that some countries may demonstrate unpredictable changes in growth rates that can not be explained within the economy and require consideration of cultural and social factors.

Turning back to the point of greater unevenness in economic development of the world regions (Table 1.3) it can be concluded that the existing inequality (following the historically formed path) will remain through the 21st century. At the present time, at GDP growth rate of 2% per year and almost the same population growth rate, the specific GDP in the poorest Africa region practically does not increase. For this region to reach the current development level of the NA region by 2100 the specific GDP growth rate should be 2.6% per year, which, based on historical experience, is an extremely high value, particularly for such a long period. If the specific GDP grows at the rate of 1% per year (which is also a rather good index) by the end of the 21st century the AF region will reach the index of US$ 4 thousand per capita. i.e. approximately the current level of Brazil.

Economic growth is associated with energy consumption growth. It is convenient to study this link based on the energy-GDP ratio $I=E/G$, its growth rate $\gamma = (1+\alpha)/(1+\beta) -1 \approx \alpha-\beta$ or coefficient of GDP elasticity of energy consumption

$\varphi = \ln(1+\alpha)/\ln(1+\beta) \approx \alpha/\beta$ (E – energy consumption, G – GDP, and α, β and γ – growth rates of E, G and I, respectively).

When GDP is expressed in comparable units (taking into account purchasing power parity) and energy consumption includes non-commercial energy (firewood, prepared by the population for their needs, etc.) with its share reaching 60% in the developing countries, the distinctions in energy intensity of the countries with low and high per capita incomes turn out to be relatively small (see Table 1.3). Thus, this index is very persistent and can be applied when analysing long-term trends in energy consumption change [44].

Relations between energy intensity and specific GDP for 31 countries, in 1985, that differ in per capita income 20 times (from US$ 1 thousand per capita in Nigeria to US$ 20 thousand per capita in Norway) [45] (Figure 1.14) can be approximated by a linear function

$$I = a+bg,$$

where $a=0.27$ tce/thousand US$, $b=0.012$ tce person/(US$ thousand)^2. It follows that per capita energy consumption $e^*=E/P$ (P is the population number) equals

$$e^*=ag+bg^2,$$

i.e. increases with growing specific GDP g faster than by the linear regularity (Figure 1.15).

The relationship presented in Figures 1.14 and 1.15 give an "instant" picture and allows comparison of energy consumption in various countries at different stages of development at the given time moment. As the statistical data analysis shows, in each individual country energy intensity first increases and then, after reaching some definite level of industrial development, starts to decrease.

Change in energy intensity for several countries during the period from 1950 (the USA, Germany), 1960 (the USSR), 1971 (Japan, China, India, Brazil, Egypt) to 1996 depending on specific GDP based on the data [22, 31, 34, 46, 47] is presented in Figs. 1.16 and 1.17. In the developed countries energy intensity in the considered period dropped, in the developing countries (except for China) it increased. Based on the tendency proved by the data analysis for several tens of countries for 1950-1988, it was suggested in [45] that there was some energy intensity level within the interval of 0.2–0.5 tce/thousand US$ to which energy intensities of different countries converge (developed countries — from above, developing countries — below).

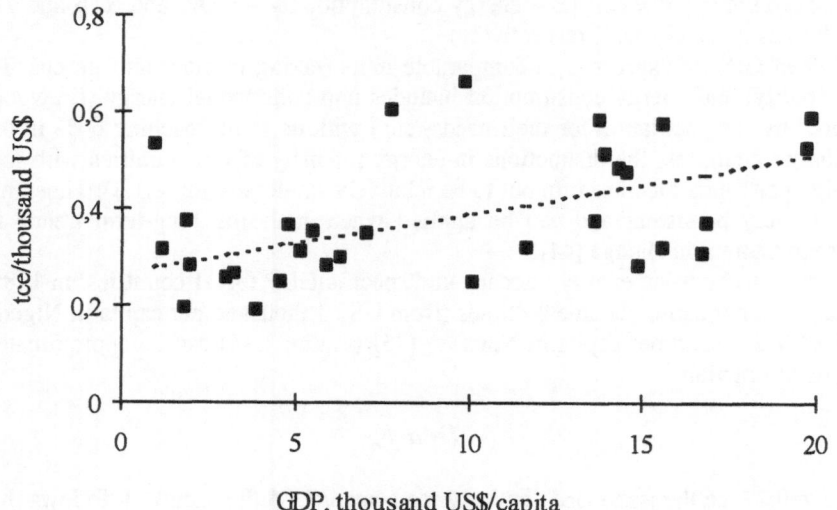

Figure 1.14. Relationship between energy intensity and GDP per capita. Dots are factual data, dotted line is an approximation.

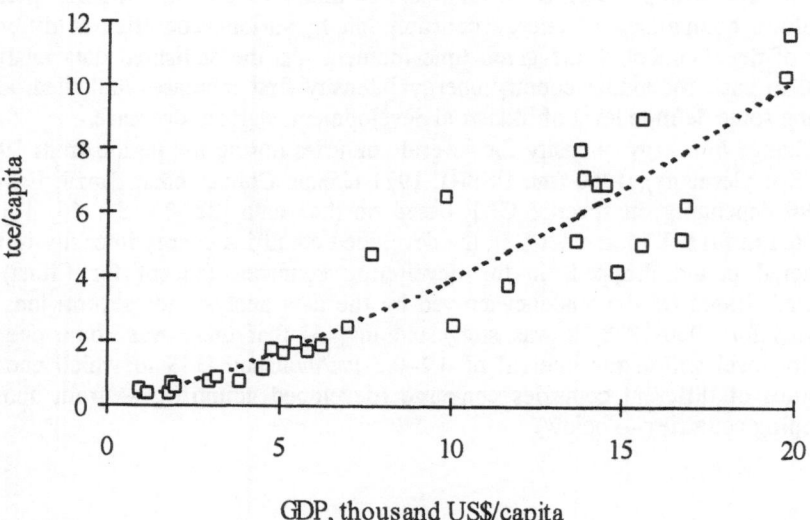

Figure 1.15. Relationship between specific energy consumption and GDP per capita. Dots are factual data, dotted line is an approximation.

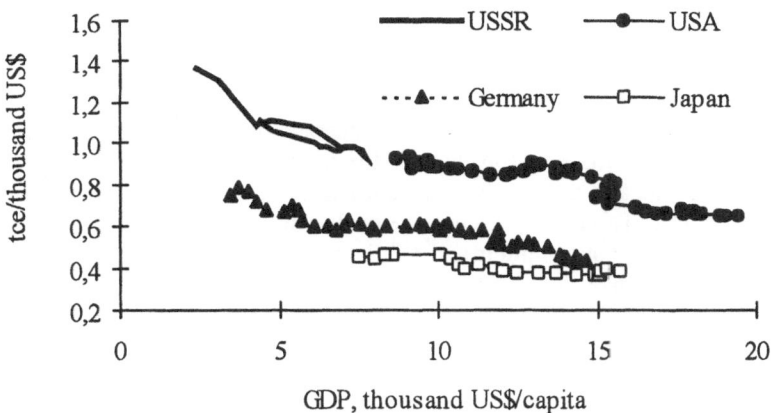

Figure 1.16. Relationship between energy-GDP ratio of developed countries and GDP per capita.

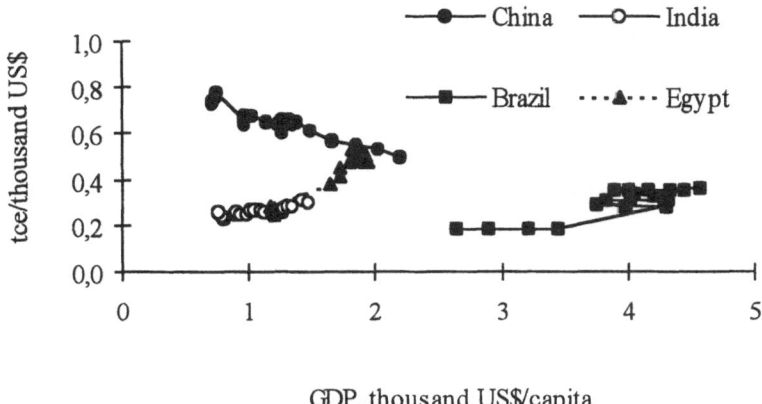

Figure 1.17. Relationship between energy-GDP ratio of developing countries and GDP per capita.

In developed countries the sharp rise in the prices of energy carriers due to oil crises in 1973 and 1979 stimulated faster decline in energy intensity. At the same time analysis of long-term trends in changing energy intensity shows that it fell in the previous period as well and maximum energy intensity was reached well before that (in 1880 — Great Britain, 1920 — the USA and Germany, 1929 — France and 1970 — Japan) [45]. In succeeding years energy-GDP ratio decreases, firstly, due to more efficient energy use (updating of production processes) and, secondly, due to structural changes in the economy in favour of less energy intensive production technologies.

1.4.   Forecasts

World economy and energy development is characterised by substantial inertia. This allows short- and long-term forecasts of the aggregated indices for countries and regions to be made and constantly corrected based on the tendencies of change in determining factors. Table 1.5 presents one of the latest forecasts (middle or the main variant) of energy consumption and GDP growth rates for the period up to 2020 made by the experts of the US Department of Energy [20], population growth rates from the data of the World Bank [21] compared to the actual indices for 1980–1990.

Table 1.5. Growth rates of population, GDP and energy consumption, % per year

Countries	Years					
	1980–1990			1995–2020		
	Population, λ	GDP, β	Energy consumption, α	Population, λ	GDP, β	Energy consumption, α
Developed	0.7	2.8	1.2	0.3	2.3	1.2
FSU & EE	0.8	2.1	1.8	0.2	2.9	1.1
Developing	2.1	3.3	4.3	1.4	4.8	3.6
World	1.8	2.8	2.0	1.1	2.9	2.1

World energy consumption will grow approximately at the same rate as in the previous period (2.1% per year). Energy-GDP ratio in the developed countries will decrease (growth rate $\gamma \approx \alpha - \beta$) by 1.1% per year; in the developing countries the tendency of energy-GDP ratio growth ($\gamma = +1.0\%$ per year) will change to its decrease ($\gamma = -1.2\%$ per year). Decline in the population growth rate will cause rise of per capita energy consumption (growth rate $\delta = \alpha - \lambda$) by 0.9% per year in the developed countries and by 2.2% per year — in the developing countries. As a result world specific energy consumption (2.3 tce/capita in 1995) will increase on the average up to 2.9 tce/capita in 2020. The structure of the consumed primary energy (Figure 1.18) will change. The natural gas fraction will increase from 21% in 1990 up to 29% in 2020, the fractions of coal and nuclear energy will decrease from 27 to 23% and from 6 to 3.5% respectively; the latter reflects the current situation with perspective of commissioning of new energy sources in individual countries and decommission of nuclear power plants with expired service life. Oil fraction as is expected will remain high enough (37–39%) and will not change significantly; the fraction of renewable energy sources will also stabilise (about 8%). Thus, renewables will play a secondary role in the world energy balance as before.

In the coming 15–20 years the following changes in energy consumption in individual sectors of the economy and final energy structure have been forecasted.

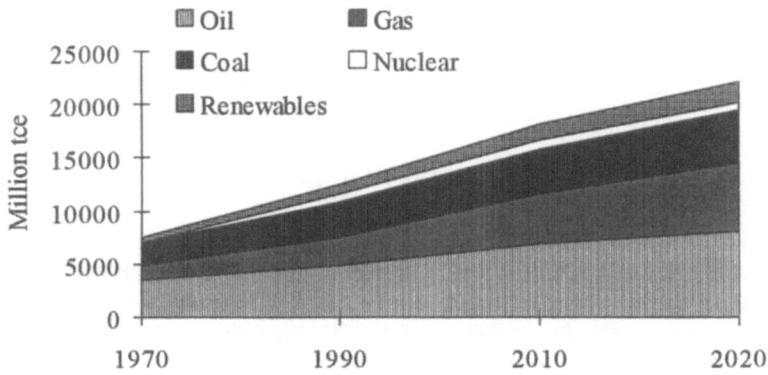

Figure 1.18. World energy consumption forecast.

1. Electrical energy fraction in the final consumption will rise, in the developed countries electrical energy will be more widely used for heating and cooling.

2. At the same time faster growth (compared to GDP growth) of electrical energy consumption observed up to the 1980s will slow down in the coming years (GDP-elasticity coefficient of electric energy consumption will be less than unity) and electrical energy-GDP ratio will decline, first of all, in the developed countries.

3. The fraction of production sphere in energy consumption of the developed countries will keep on declining and the transport fraction will grow. The fraction of non-production sector will remain almost constant (before the early 1990s it was gradually increasing and then began to decrease slowly).

Thus, short- and medium-term forecasts of world energy development with regard to the current (and constantly changing) situation on world energy markets show that in the coming years (decades) fossil fuel and conventional energy technologies will be the basis of the world energy as before. However, if we consider more remote future (50 years and more) more significant changes in energy structure are likely due to depletion of the cheapest categories of energy resources.

Currently economic efficiency of energy technologies depends on changing energy resources prices. Formation of oil price (determining other prices due to substitution of energy carriers) is affected by   changing explored reserves, production volumes and cost, demand, policy of countries-exporters and many other factors. Consideration of all these factors is practically impossible in the long-term studies. However, in the future the role of production cost in price formation will be larger due to limited fossil fuel resources.

At present the total world fossil fuel resources may be thought of as known (with a certain degree of assurance). Based on the results of the study [48] world oil resources (from cheap oil of the developed fields with a production cost up to US\$ 60/tce to hypothetical resources of unconventional oil (unknown so far) with a production cost up to US\$ 780/tce) amount to 3.8 trillion tce, natural gas — 29 trillion tce (up to US\$ 700/tce) and coal — 8.9 trillion tce (US\$ 5-180/tce).

Knowing distribution of these resources based on production cost it is possible to determine future fuel price against its consumption rates: with the production growth rate $1+\alpha$, i.e. when $E(t) = E_0(1+\alpha)^t = E_0 \exp(\alpha^* t)$, the i-th cost category of resource ΔE with production cost from C_i to C_{i+1} is consumed in time

$$\Delta t = (1/\alpha^*)\ln[1 + \alpha^* \, \Delta E/E_0],$$

where $\alpha^* = \ln(1+\alpha)$ and fuel cost increases up to C_{i+1}.

Let us consider two variants: 1) oil, natural gas and coal production increases in accordance with the forecast [20] (see Figure 1.18), i.e. with the growth rates 1.8, 3.3 and 1.7% per year; 2) current energy structure is retained and energy consumption growth rate decreases from 2% per year, existing and forecasted for the next 20 years, to 1.5% per year. For simplicity sake assume that the fuel price is proportional to its production cost and averaged world production costs and prices [49] in 2000 are equal to US\$ 40 and 90/tce for oil and gas, US\$ 20 and 40/tce for coal respectively.

The calculation based on these data (Figures 1.19 and 1.20) show that there will be no significant change in fossil fuel prices up to 2020–2030 and hence, the existing energy structure will not require any fundamental reformation. Oil and natural gas demand growing in accordance with the forecast [20] can be met by using cheap categories of resources. At the same time, by the middle of the 21st century even in the variant of slow production growth, the oil price will increase 3 times, which will influence its competitiveness as compared to the other energy carriers (including synthetic liquid fuel produced from coal). Towards the end of the century only one kind of cheap fossil fuel — coal — will remain.

The relation between reserves and current production is often used to characterise provision of a country or region with energy resources. For the majority of oil producing countries this relation is in an interval of 30–50 years [49]. However, this does not mean that oil resources will be completely depleted already in some decades since only proved (explored) reserves are taken into consideration in the given case.

The above estimations show that even at very high (hypothetical) rates of fossil fuel consumption its resources (those explored to date and forecasted — to be explored in the future) will be sufficient for the whole 21st century. Hence, we should deal here with fossil fuel provision as well as with the time when fuel becomes expensive as to essentially restructure energy use and to use mainly the renewable kinds of energy, nuclear and thermonuclear energy.

Concerns that scarcity of resources will have effect already in the coming years, encouraged (especially after the oil crisis of the 1970s) special studies on long-term prospects of world energy development. Let us dwell (very briefly) on the two most comprehensive and basic studies carried out by the International Institute for Applied Systems Analysis (IIASA, Vienna, Austria) in 1973–1979 headed by W.

Häfele [2] (the project IIASA-1) and in 1993–1998 under the direction of N. Nakicenovich [17,33] (the project IIASA-2).

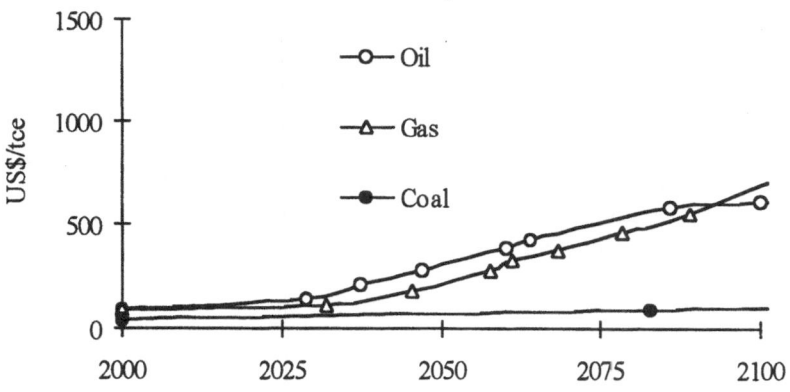

Figure 1.19. Fossil fuel prices at its consumption growth rates: oil –1.8, gas – 3.3 and coal – 1.7% per year.

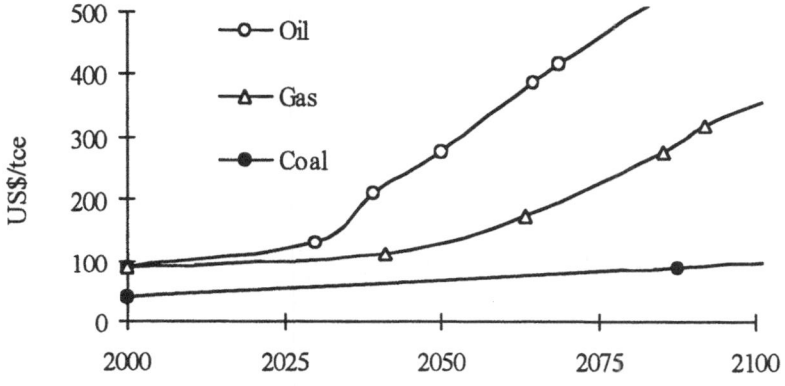

Figure 1.20. Fossil fuel prices at oil, coal and gas consumption growth rate 1.5% per year.

From general considerations it is clear that in the remote future mankind should turn to using inexhaustible energy resources (solar, nuclear, thermonuclear energy, etc.) and clean energy technologies. The main problem raised in the project IIASA-1 dealt with the terms and ways of transition to a new energy structure. It turned out that in the following (after 1980) 50 years, i.e. up to 2030, fossil fuel will still be prevailing and insufficiency of resources will show up at a later period.

The next project (IIASA-2) was implemented based on updated information. Here more attention was given to forecasting by the use of mathematical models of scientific and technological progress in the energy sector and its influence on development. Consideration was given to a wide spectrum of possible scenarios for the economy and energy development varying in economic growth rates, development and introduction of new technologies, access to different energy resources as well as political aspects (international energy resources trade, exchange of technologies, settlement of environmental protection problems). The obtained results confirmed the conclusions of the previous study on the important role of fossil fuel, at least up to the middle of the 21st century, and showed a principal possibility of creating vastly different world energy structures meeting different requirements and constraints — technical, environmental, political.

Figure 1.21 presents the results for some scenarios considered in the projects IIASA-1 and IIASA-2 compared to the latest forecasts of the US Department of Energy [20] (DOE).

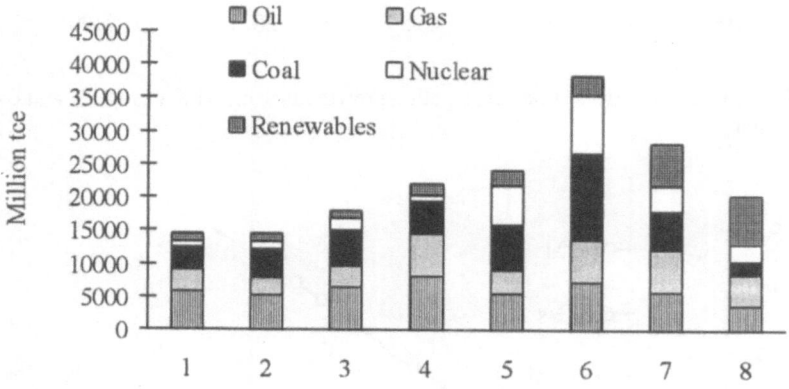

Figure 1.21. World primary energy production. 1 – 2000 (DOE); 2,3 – 2000 (IIASA-1), low and high scenarios; 4 – 2020 (DOE); 5,6 – 2030 (IIASA-1), low and high scenarios; 7,8 – 2050 (IIASA-2), scenarios B and C2.

It is evident that in 1980–2000 the low scenario of IIASA-1 was realised (energy consumption growth rate amounted to about 2% per year), however the nuclear energy fraction turned out to be lower than the forecasted one due to non-economic constraints which could not be foreseen in 1979. In the future development is likely to follow the scenario of lower energy consumption with a lesser fraction of nuclear energy and larger fraction of renewable energy sources as is evident from the results of IIASA-2 for 2050.

The studies of long-term prospects of world energy development that have been carried out by now give a good methodical and information basis for further studying many unclear problems, which will be addressed in the next chapters of the monograph.

Chapter 2

METHODOLOGY OF STUDIES

2.1. Energy and sustainable development

The necessity of radical and sufficiently rapid (on a historical scale) changes in the options for social and economic development of mankind and for transition to "sustainable development" is most comprehensively substantiated and reflected in the resolutions of the UN Conference in Rio de Janeiro in 1992. These changes will obviously apply to the energy sector, influence its development, and place new requirements upon it, which should be taken into consideration in devising a new methodology of energy investigations and forecasting.

The main conclusion of the UN Conference in Rio de Janeiro was that mankind can not continue developing in the traditional way that is characterised by irrational use of energy resources and increasing negative impact on the environment. If the developing countries follow the way by which the developed countries reached their well being, a global ecological catastrophe will be inevitable. The world community should pass to a new model of development, called sustainable development.

The Rio Declaration on the environment and development and "Agenda for the 21st century" [50] adopted at the mentioned UN Conference reveal the essence and goals of the "sustainable development" concept, as well as the anticipated interrelations between national and global interests, the role of the state and different classes of population, problems arising in transition to sustainable development, etc. The Rio Declaration, for example, contains 27 recommended principles, including a number that outline a notion of sustainable development:

— recognition of a priority for social factors ("the right of people to have healthy and fruitful lives in harmony with nature is a principle focus of attention");

— recognition of inseparability between the processes of development and environment conservation ("environmental protection should become an integral component of the development process and can not be considered in isolation from it");

— recognition of interests of future generations ("the right of development should be implemented so that requirements for development and environmental protection of present and future generations are equally fulfilled");

— provision of social equity both in each country and among countries ("decrease of the gap in the living standard of people on the planet, elimination of poverty and misery");

— recognition of the states' responsibility to their peoples and the world community for activity aimed at development and environmental protection.

It should be emphasised that the problem of sustainable development is based on the objective necessity (as well as the right and inevitability) of socio-economic development of the third world countries. The developed countries could apparently content themselves (at least for some time) with their achieved levels of wellbeing and consumption of the planet's resources. However, their responsibility is not only to conserve the environment, but at the same time to enhance a socio-economic level of developing countries ("South") and to bring it closer to the level of developed countries ("North").

The Conference in Rio de Janeiro was followed by a number of conferences connected with the concept of sustainable development in some way or another. They are: Conference on the world population in Cairo in 1994, World Congress on women's status in Beijing in 1995, Summit of cities in Istanbul in 1996, Special 19th Session of UN General Assembly in June 1997, Conference on global warming in Kyoto in 1997 and the later conferences of participants of the UN Framework Convention on Climate Change in Buenos Aires in 1998 and in Bonn in 1999.

In Russia studies in the indicated direction have been under way rather intensively in recent years. They are most completely reflected in the encyclopaedic monograph [51] that was written with participation of about 70 scientists and experts, including more than 20 Academicians and Corresponding Members of the Russian Academy of Sciences. The monograph deals with social-political, economic, spiritual-moral, environmental and other aspects of transition to sustainable development, specifically in Russia. It also describes the conception of transition of the Russian Federation to sustainable development that was approved by the President of RF in April, 1996 and comments on it. Approaches to elaboration of the National Strategy of sustainable development for Russia are suggested. Insufficient (for the current conditions) activity of the governmental authorities of RF on practical transition to sustainable development, especially in social, economic and moral problems, is emphasised.

Transition to sustainable development raises a set of diverse problems (their majority is presented in "Agenda for the 21st century"). Financial and technological assistance of the developed to the developing countries and establishment of quantitative constraints and quotas on environmental pollution and on the use of non-renewable natural resources are among the most difficult problems. Without such assistance transition to sustainable development is impossible. However, solution of the first problem will require radical changes in the consciousness of people in the developed countries and elaboration of new principles of international relations. Difficulties of the second problem are caused by both a wide variety of indices describing the state of environment and the necessity for scientific substantiation of their quantitative values and the subsequent co-ordination of their approved (directive) values at the level of governments and international organisations.

A lot of indicators have been suggested that characterise sustainable development [52–55]. The main difficulty here is to define their "threshold" (minimum or maximum admissible) values, at which achievement the development could be reckoned as sustainable. For example, the specific (per capita) gross domestic product (GDP) of individual countries and regions, the specific consumption of main kinds of goods and services (including energy) by population, the indices of environmental pollution (for example, CO_2 emissions), etc. might be referred to such indicators. However, there is no clarity as yet concerning the maximum permissible indices even for CO_2 emissions and concentrations despite extensive investigations carried out in the world. And the CO_2 problem is not the only one. Even if it is solved successfully, the necessity of transition to sustainable development will remain.

There are other, difficult to estimate, factors that could lead to global catastrophe. They include: wars of mass destruction (especially with nuclear weapons) that are caused by the socio-economic inequality between "North" and "South"; irreversible pollution of the world's oceans; physical degradation of populations, caused by disease, hunger, and unhealthy modes of living. It is necessary to apply both quantitative indices and qualitative judgements in our studies of sustainable development. In particular this is needed to generate a variety of scenarios reflecting expected (possible) conditions of that development.

The importance of action by regional (a group of countries) and global entities considerably increases. It is evident that each individual country is directly involved in elaboration and implementation of its own program of sustainable development. However, the objectives and some parameters (indices) should correspond to international agreements made at the global or regional level. For the world as a whole, and then for its regions and individual countries, one should establish environmental constraints and evaluate consequences of their introduction, elaborate a general policy of using non-renewable resources of the planet, determine required and feasible scales of assistance of the developed to the developing countries and realistic rates of bridging the gap in their living levels.

On the whole, transition of mankind to sustainable development seems inevitable. Therefore, studies on the ways of such transition for the energy sector as well are undoubtedly needed, particularly in making long-term forecasts of energy development at the country, region and world levels.

Now let us consider *energy requirements for sustainable development*. It has long been said that the energy sector is about to enter, or already has entered, a transition period. It is difficult to refute this contention. However, we need to specify what we mean by transition, as its concept changes over time.

In the 1970s, during the energy crisis, consideration was given to *transition from using limited fossil fuels to virtually inexhaustible or renewable energy resources*. Special emphasis was placed on nuclear energy with breeder reactors. It was proposed to create nuclear "islands" [56], the global energy system [57], etc. The most comprehensive studies on energy problems were conducted at that time at IIASA [2].

Later on, in the 1980s and 1990s, particularly great attention was paid to energy-related environmental problems. Interesting conceptions were proposed for integrated environmentally clean energy systems [58] based on the use of fossil fuel as well as nuclear energy. Synthetic energy carriers (methanol, synthesis-gas, hydrogen) and the environmentally clean technologies of their production and utilisation are of particular significance in these systems. It should be emphasised that the term "environmentally clean" can not be interpreted as "absolutely clean". More likely it means environmentally acceptable or relatively clean energy technologies and systems. In recent years a wide range of works have been devoted to the problem of limitation (reduction, stabilisation) of CO_2 emissions of global character (see, for example, [59]).

In this context the transition period came to be perceived as *transition to environmentally clean energy systems* [58,61 , etc.] assuming that they will be based on inexhaustible energy resources. Electricity and hydrogen, which are environmentally clean and handy energy carriers, might underlie such energy systems. And they should be produced from such virtually unlimited primary energy resources as solar and nuclear energy (the use of breeder reactors on uranium and thorium or fusion reactors). The problem is that such an "electrical-hydrogen" system will call for extremely large financial, labour and material resources and, for the majority of regions, its creation is feasible only toward the end of the 21st century or even later. Therefore, mankind would have to use cheaper fossil fuel for many decades to come, bearing additional expenses for cleaning the products of its processing and combustion.

To study the ways of creating environmentally clean energy systems at the national, regional and global levels, methodological approaches and a great number of mathematical models were worked out (see, for example, [60] and also [7, 61] written by the authors of this book). Note that in the majority of works concerning CO_2 emissions basically environmental aspects are investigated, though some issues of sustainable development are also dealt with.

Now let us to return to *energy for sustainable development*, transition to which should be considered with current views on the situation in the world. One can see that the requirement of inexhaustibility of energy resources used and the environmental cleanness envisaged in the concept of environmentally clean energy systems satisfy two most important principles of sustainable development (observance of interests of future generations and conservation of the environment). Besides, it was always supposed that energy development should be considered in close interrelation with the economy (of the given country or region).

Analysis of the other principles and specific features of the concept of sustainable development allows the conclusion that at least two additional requirements must be imposed upon the energy sector (Figure 2.1):

— provision of energy consumption (including energy services for population) that will be no lower than a certain social minimum;

— co-ordination of development of the national energy sector (and the economy as well) with its development at regional and global levels.

The fourth requirement stems from the principles of priority of social factors and assurance of social equity. For realisation of the right to healthy and fruitful lives, bridging of the gap in the living standard, elimination of poverty and misery, establishment of a certain living wage, while including satisfaction of the minimum necessary energy demands of the population and the economy, should all be guaranteed. Clearly this requirement concerns first of all the developing countries, determining the lower limit at which their development may be considered sustainable. However, it also applies to the developed countries, if the necessity of energy conservation and reasonable limitation (or even reduction) of energy consumption is taken into account. In this case the minimum energy consumption that will provide an admissible level of living and economic development is implied.

Advancement of such a requirement poses new, rather challenging problems on quantification of the needed minimum, on the one hand and on its realisation, on the other. Satisfaction of the minimum energy demands of the developing countries will be, in particular, one of the directions of rendering assistance by the developed countries.

ENERGY FOR SUSTAINABLE DEVELOPMENT	
1. Use of inexhaustible primary energy resources or those sufficient for very remote future (several centuries) 2. Provision of environment conservation (use of environmentally clean technologies and secondary energy carriers) 3. Co-ordination (proportionality) of energy and economy development	Environmentally clean energy sector
4. Provision of energy consumption (including energy services to population) that will be no less than a certain minimum 5. Co-ordination of energy development at national, regional and global levels	

Figure 2.1. Requirements to the energy sector.

The fifth requirement is caused by the global character of the threatening ecological catastrophe and the need for co-ordinated actions of the world community on elimination of this threat. Even countries that possess sufficient energy resources (Russia, for example) can not plan independently their energy development because of the necessity to take into account global and regional environmental and economic constraints. This is most typical of the developing countries and regions, to which the financial and technical assistance by the developed countries should be rendered and distributed in some way among them. The scale of assistance is also important for the developed countries, since it leads to reduction in financing of their own economy and energy development.

As to the methodology, analysis of ecological as well as economic and social factors, and also the economy and energy development at the regional and global levels, should be carried out to take into consideration specific features of sustainable development in energy studies. It means the necessity to devise new indices, improve and apply new mathematical models, increase the number of scenarios of external (with respect to energy) conditions, etc.

2.2. General approach, composition and sequence of studies

The methodology of studies of long-term prospects in energy development at the global level was formed and improved in the course of study of a widening scope of tasks and problems. Initially the studies were of a technological nature and were aimed at estimation of the *effectiveness, role and possible scales of using new energy technologies and systems at the global level.* These are technologies for production of electrical energy and synthetic fuel, use of nuclear and renewable energy forms, the Lunar power system, the technologically unified world energy system, etc. Account was also taken of global constraints on CO_2 emissions, development of new technologies and some others.

Later on the range of tasks was widened and included, in particular, such as:

— analysis of the world energy balance in terms of the possible role (time of depletion) of cheap non-renewable resources (oil, natural gas, uranium–235);

— determination of the critical term for introduction of new energy carriers (synthetic liquid and gaseous fuel);

— analysis of regional peculiarities in energy development and possible scales of export/import of energy carriers;

— analysis of demands for capital investments (and total expenditures) in energy development.

Such studies resulted in revealing long-term trends in world energy development, technological progress being taken into consideration. These trends are of interest in many respects, including decisions made on rational directions of technological progress in the energy sector, which require a long lead- time.

In recent years, as experience in studies was accumulated and problems of transition to sustainable development arose, the authors have set two additional objectives:

— to formulate a concept of energy for sustainable development;

— to generate maximum "plausible" forecast of energy development for the world and its regions (in contrast to variants usually elaborated by the principle "what will happen, if …").

Thus, the presented methodology is intended for solving a sufficiently wide range of problems and the authors have tried to make it more universal. From the outset it was based on the methodology of long-term energy studies at the global level that was developed at IIASA [2] and includes the following components (Figure 2.2):

1) application of mathematical models;

2) use of the scenario approach to take into account information uncertainty on the future conditions of energy development;

3) division of the world into large regions in accordance with specific principles;

4) analysis and preparation of the information on energy demands, resources and technologies;

5) sizing up the existing environmental, economic and technological constraints on energy development (political and other similar factors were not considered).

The computations are performed on mathematical models and the results obtained are analysed to make proper conclusions and recommendations. IIASA's studies were versatile (multi-factor), comprehensive and had an interdisciplinary nature. The methodology of studies applied and described in the book differs in some aspects from IIASA's methodology (develops it) and the approaches used by other authors.

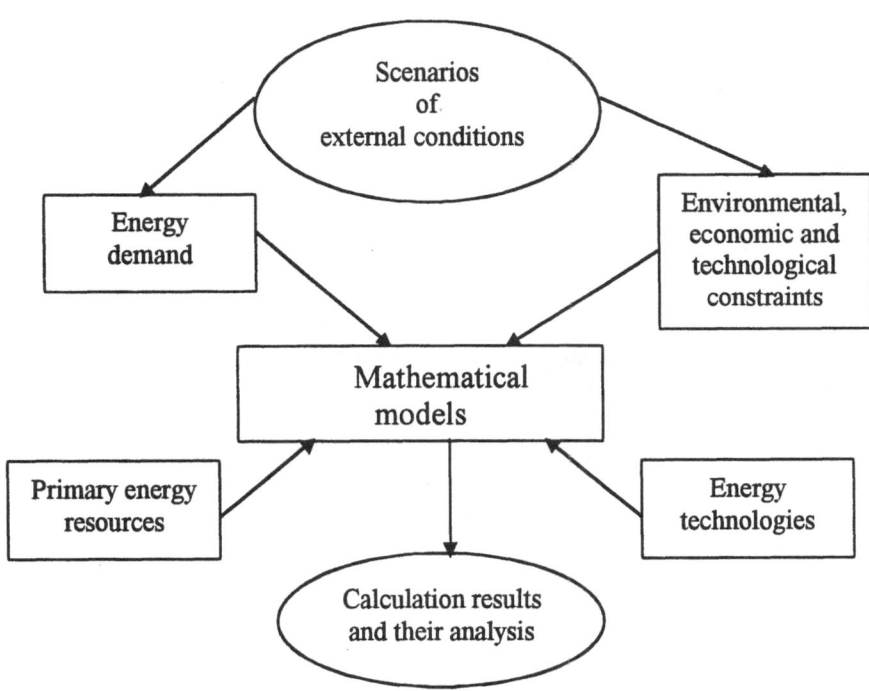

Figure 2.2. General scheme of studies.

The *optimisation global multi-regional mathematical model GEM-10R,* whose type and structure determined key features of the methodology, is the pivot of the studies described. Its mathematical formulation is given in the next section. Here we underline the most interesting methodological peculiarities of the model (capabilities, merits and usually inevitable shortcomings).

1. The GEM-10R model belongs to a class of mathematical models that are based on the "bottom-up" approach and is intended for determination of the optimal *technological structure* of the world energy system with specified energy demands. The aggregation level of technologies, energy forms and territories was chosen based on the studied time horizon (1990–2100). Hence, calculations make a forecast (under some or other external conditions) of the *long-term global trends* in energy development that can be extended to the level of local problems or short-or medium-term future only with great care.

Computations with the given energy consumption do not take into consideration its dependence on energy cost (in distinction to the macroeconomic "top-down" models). Therefore, different consumption levels were chosen and then the computation results were studied for consistency by the macroeconomic model describing the energy-economy interrelations.

2. The model optimises an energy structure of the world as a whole with division into several (as a rule 10) regions, allowing one to determine a mutually co-ordinated structure of regional energy systems with regard to the export/import of energy carriers. This is its great advantage in comparison to the regional models (MESSAGE, MARKEL, EFOM, etc.) used for energy studies at the global level, when there is a need to specify an interregional exchange of fuel and energy in some way (usually by expert judgement).

3. The model is static. The computations are performed at steps of 25 years and the technological structure of the energy sector is renewed almost completely every time. Such an approach conforms to the fact that the service life of the majority of technologies is close to 25 years. Only some "long-lived" objects and technologies, hydropower plants as an example, are "accumulated" and transferred from one step to another with the remaining non-renewable energy resources and other analogous data, making the model quasi-dynamic.

The static nature of the model is a certain disadvantage compared to dynamic models. In particular, sometimes sudden changes in the regional energy structure are observed when passing from one 25-year period to another. They should be smoothed in the repeated calculations by imposing additional constraints. At the same time the static nature of the GEM-10R model opened up considerable opportunities to investigate a long-term time horizon (100 years and more), since it allows a consecutive and equally accurate optimisation of the energy structure for both the nearest and remote 25-year periods. When the dynamic models are applied for optimisation simultaneously on the whole studied period, its end is lost ("disappears") at a duration of more than 25–30 years due to cost discount.

4. The model considers (and sets as initial information) demands for *final* energy (electrical, thermal, mechanical, chemical). Therefore, different approaches to energy consumption forecasting as compared to the traditional determination of demands for electrical energy, solid, liquid and gaseous fuel (i.e. in fact the primary energy demands) are needed. However, at the same time this offers the possibility to model the whole energy chain from primary energy production to final energy consumption, choose technologies of final consumption and determine primary energy expenditure (rather than set its demand). This characteristic of GEM-10R is methodically its further advantage.

5. The objective function of the model is the discounted total expenditures (investments and operating costs) on regional energy systems (for the whole technological chain) and the interregional transport of energy carriers for the 25-year period. Hence, optimisation is performed for the "whole world".

Such an approach can provoke an objection, because it does not take into account contradictory interests of individual regions, possible economic and political conflicts that are described by the game and market models. This should be taken true for a short-term time interval (5–10 years), when the fuel prices and the volume of fuel trade are formed in response to the specific state of the market and the policy of exporting (or importing) countries.

For the long-term future, the economic factors become crucial. The economically most attractive energy resources and technologies will find their way despite possible variations in the market state and political barriers. Therefore, optimisation by economic criteria reveals those long-term trends in world and regional energy development that are most reliable and probable in real situations.

Theoretical analysis shows that the minimum expenditures solution of the global optimisation problem coincides with maximum profit solutions of local problems, if appropriate global constraints are taken into account. In this context the GEM-10R model simulates to some extent competition of regions and energy technologies, i.e. the market.

6. In the course of model application the authors required the highest possible differentiation of primary energy resources (non-renewable and renewable) with respect to cost categories. At first the resources were divided into 2-3 categories, later 4-6 and presently 7-9 categories of cost. This fact makes it possible to account for a non-linear dependence of the cost of resources on the scale of their use, to determine more exactly dynamics of depletion of cheap resources and substitution (competition) of different resources, and to take into consideration their very expensive (sometimes speculative) reserves.

7. The global model allows for global constraints, for example on CO_2 emissions, and determines their optimal distribution among the world regions. It also allows regional constraints (environmental, technological and economic) to be introduced in parallel.

Other mathematical models, along with the global model, were applied in the studies to some extent (Figure 2.3). However, they do not form a single computation complex and are used in different combinations depending on the objectives and peculiarities of the particular study.

The GEM-10R model with division of the region into smaller groups of countries was used as the *regional energy model*. Specifically, studies for the time horizon to 2050 were performed for the former USSR region with its division into seven parts (in particular, Russia was divided into four parts). The results obtained were more precise than with application of the global model, especially for nuclear energy development. In calculations on the global model, such a huge region as the USSR was represented by "one point" and nuclear energy turned out to be inefficient (was substituted by Siberian coal). However, factually it is economically efficient in the European part and in the Far East of Russia and also in the European countries of CIS.

The models of *energy plants* were constructed to determine technical and economic indices of new (perspective) technologies of coal conversion [62]. Then these indices were used in the global and regional energy models. The models of wind/diesel plants (systems) are an example of such models [63]. More sophisticated system models may be needed to study prospects in development of specialised (branch) energy systems (electric power, heat supply, etc.) as well as such radically new systems as integrated energy systems [58, 61, 62], the Lunar power system (see Section 5.5) and so on.

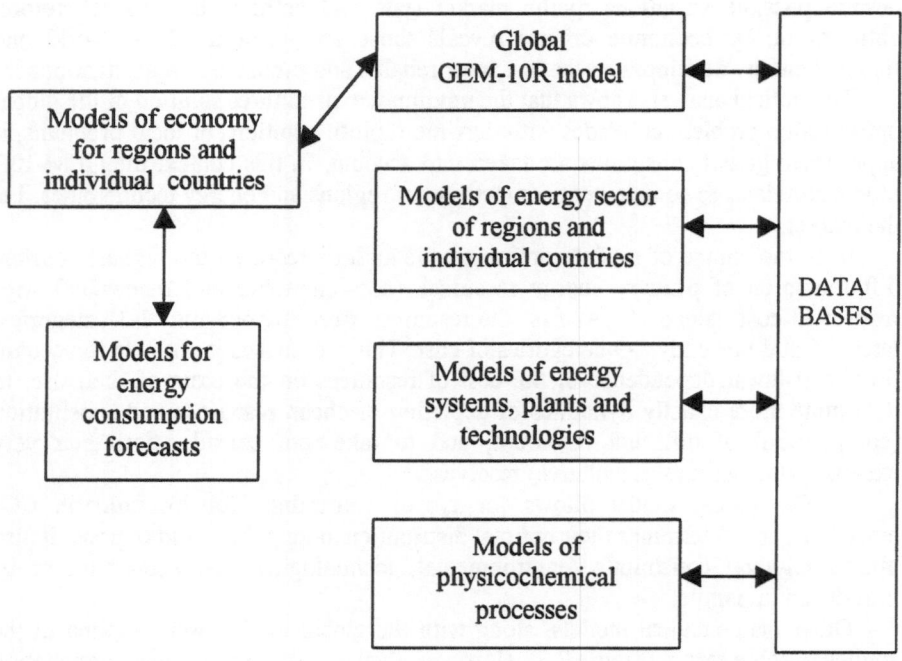

Figure 2.3. Models applied for investigation of long-term prospects in energy development.

Among the models of *physicochemical processes*, those that were worked out and applied to the greatest extent were thermodynamic models which were used to study new energy technologies, determine their expedient characteristics, and assess their possible (thermodynamic) limits for improvement [61]. These models were used to investigate coal conversion and combustion processes, in particular to determine (minimise) emissions of harmful matters (ecological characteristics of technologies).

The economic models were applied for studies on the economy-energy interrelations (basically in the same manner as in IIASA's methodology), namely for:

— determination of GDP dynamics in the regions and estimation of the effect of energy expenditures on it;

— estimation of energy consumption depending on the energy cost;

— determination of a fraction of expenditures on energy development in GDP of the regions, etc.

A model that is similar to that described in [64] is dealt with in Chapter 9.

The models of *energy consumption forecasting* have not yet been developed by the authors. This is planned further on as applied to the methods of determining minimum energy demands that are described in Chapter 3.

Let us briefly consider specific features of other components of the methodology of studies presented in Figure 2.2.

Scenarios of external conditions for energy development are generated based on specific goals of studies. Thus, when studying the impact of constraints on CO_2 emissions or a moratorium on nuclear energy development, the pairs of scenarios, in which such constraints are present or absent, were formed. In other respects (for example, in energy consumption levels) the scenarios remained identical, allowing the paired comparison of computation results, including expenditures on energy development. In this case the "price" of imposing the corresponding constraint could be determined.

Scenarios simultaneously differing in several characteristics (energy consumption levels, environmental and other constraints, etc.) can also be generated, when needed (or expedient). In particular, it concerns "plausible" scenarios for making forecasts on energy development in the 21st century that will be described in Chapter 6.

The forecasts of *energy demands* are of considerable importance in the described methodology. One of the peculiarities that was mentioned above is to determine demands for *final* energy forms. At first the forecast is made for specific (per capita) consumption in different regions, and then the absolute consumption of the indicated four forms of final energy is determined from the forecasts of population number in the regions.

The energy consumption forecasts have been corrected in the course of multiyear studies. A latest forecast was elaborated within the framework of the IAEA project, mentioned in the Preface, with participation of other executors of the project. The results obtained for this forecast convinced the authors that it was hardly real (too high), especially if account is taken of conditions and requirements for transition of mankind to sustainable development. Therefore, we decided to elaborate a forecast of the minimum energy consumption that is necessary for the developing countries or sufficient (with regard to energy conservation) for sustainable development of the developed countries.

The authors realise well the complexity of such a problem, when even the notion "minimum energy consumption" itself is difficult to formulate, and besides it differs for the developed and the developing countries. In fact this requires a deeper analysis of social needs and economic conditions, consideration of distinctions in the mode of population life in different regions and countries, etc. This in turn will call for lengthy and versatile investigations.

The thermodynamic analysis was suggested as a possible way to determine minimum energy consumption. It implies the analysis of idealised processes of production of energy-intensive goods and services with determination of the theoretical energy efficiency or theoretically minimum (according to the laws of thermodynamics) specific energy consumption for production of these goods (or energy services). This efficiency or specific energy consumption characterises theoretical limits in the improvement of energy consumption processes that were studied. [65, 66].

The actually achievable efficiencies will surely be lower (and the specific energy consumption will be higher) than the theoretical ones. Therefore, a need arises for the analysis of observed dynamics and forecasts of efficiency improvement for processes and transition to new, more perfect technologies. In parallel there is a need to forecast specific (per capita) consumption of energy-intensive products in different world regions (and population in the regions). Despite the considerable labour input of such an approach the authors believe that it will furnish a means for more valid determination of the minimum energy demand in industry.

Besides, two additional approaches to forecasting the minimum energy consumption were tested. The first, similar to the one used in [5], was applied to the developing countries assuming that they will come close (with a certain time lag) to the living standard in the developed countries, however with regard to technological progress achieved by now (or by the forecasted time horizon). The second (more "traditional") approach is based on the hypotheses of the regional GDP growth and its energy intensity decrease, simulation of dependence of the GDP growth rates on its specific (per capita) values, etc. The three approaches applied to forecast minimum energy consumption are described in the next chapter.

Primary energy resources, as was noted, are divided into several (seven – nine) cost categories based on the accessible publications, especially [48, 67]. Resources of wind and solar energy for different cost categories were estimated by the authors.

Highly diverse *constraints on energy development* were determined depending on their nature and changed with accumulation of the experience of studies. At first the most important constraint on global CO_2 emissions was taken equal to 1990 emission levels. Then it was relaxed in accordance with work [68]. However, even in this form the constraint seemed too rigid and hardly feasible. In the latest calculations on GEM-10R, in the "plausible" scenarios as well, it was taken at a level of 40–55 Gt/year for CO_2 (or 11–15 Gt/year for carbon) for the second half of the 21st century (considering work [69]).

Some analysis was made to determine regional constraints on emissions of sulfur dioxide (SO_2), nitrogen oxides (NO_x) and particulates — soot, dust, etc. These constraints were imposed from the admissible concentrations of harmful substances on densely populated territories of each region. For large regions such territories make up only several percent of their total area.

It should be emphasized that emissions of SO_2, NO_x and particulates and those of CO_2 exhibit close correlation. Introduction of constraints on CO_2 emissions reduces emissions of other harmful substances and vice versa. Introduction of both types of environmental constraints leads to more precise determination of the admissible fossil fuel consumption in different regions.

Constraints on development or use of different energy technologies were established primarily from the analysis of potential terms and rates of their introduction. Constraints on the use of nuclear technologies in different regions that are foreseen in the "plausible" scenarios were put in terms of potentialities to guarantee their security, skilled operation and some other considerations (see Chapter 6).

The economic constraints (on capital investments in and total expenditures on the energy sector) have not been imposed as yet, though in the GEM-10R model this fact is envisaged. Analysis of expenditures and energy − economy interrelations was made after calculations on the model (see Chapter 9).

A mix of *energy technologies* was determined in terms of the GEM-10R structure (see Section 2.3), on the one hand and from the analysis of all accessible information on the expected technological progress in the energy sector.

For indirect consideration of potential "surprises" in technological progress, the energy production technologies include two "marginal" technologies which are very expensive but have practically unlimited scales for application. The former comprises fast breeder reactors operating on the "waste" uranium-238. Other new nuclear technologies (thermonuclear reactors, thorium reactors, etc.) are supposed to be hardly cheaper than breeders. The latter is the Lunar power system (see Section 5.5), whose scales can potentially be very large (tens of thousands of gigawatts). The availability of this, also very expensive, technology allows one not to consider other exotic renewable energy sources.

Analysis of the computation results on mathematical models (a closing point of the methodology) has no clear algorithm and is made with regard to the objectives of a particular study. Following the objectives stated in the Preface, such an analysis is made in Chapters 7−9. Besides, Chapter 10 describes additional results on the problems of technological progress.

2.3. Mathematical model of the world energy system GEM-10R

2.3.1. Areas of application

1. Global energy studies for the long-term perspective:
— analysis of the world energy balance in terms of the future role of fossil fuels, nuclear and renewable energy sources in the energy sector of different world regions;
— determination of the economic feasibility, scales and terms of using new energy carriers;
— study on the long-term prospects for using new energy technologies (terms, scales and sphere of application);
— study on the impact of global and regional constraints (environmental, including CO_2 emissions, financial, nuclear moratorium, etc.) on energy development in regions and the world;
— study of the role of particular regions in formation of the world energy market;
— optimal distribution of quotas on CO_2 emissions among the regions.
2. Determination of requirements to the energy sector at a transition of the mankind to sustainable development.

3. Solution to some strategic problems of energy development in Russia:
— energy independence;
— export/import policy;
— ecological security;
— export of energy technologies, etc.

The model is implemented as the computing system *GEM-10R (Global Energy Model: 10 World Regions)*.

2.3.2. Mathematical description

As was noted, the model is linear, optimisation, multi-nodal, quasi-dynamic [8, 70]. The number of nodes $r \in R$ is equal to 10 and corresponds to the accepted regionalisation of the world. For each region *r*, given are:

1. *List of primary energy resources* (I_1) comprising in a general case oil, natural gas, coal, nuclear fuel, hydropower, biomass, wind, geothermal, solar energy and power from space (from LPS). Each resource is split into several cost categories and constraints on their mining (production) are assigned for each time period *t* — $b_{tri}, i \in I_1$.

2. *List of secondary energy carriers* (I_2): liquid hydrocarbons of two kinds: a) light ones which are used as motor fuel (gasoline and diesel fuel) and feedstock for the chemical industry, b) heavy ones (fuel oil), as well as methanol, substitute natural gas (SNG) and hydrogen.

3. *List of final energy forms* (I_3): electrical, thermal, mechanical and chemical. Demands for the indicated final energy forms are assigned for each time period *t* and region *r*, $b_{tri}, i \in I_3$.

4. *List of typical power consumption modes* (I_4) with their numerical values — $b_{tri}, i \in I_4$.

5. *List of harmful substances* emitted into the environment: SO_2, NO_x, particulates and CO_2 (I_5). The corresponding constraints are assigned for each substance — $b_{tri}, i \in I_5$.

6. *Constraints on investments* — $b_{tri}, i \in I_6$.

7. *List of technologies*: a) mining (production) of primary energy resources — J_1, b) import/export of energy carriers — J_2 / J_3, c) production of secondary energy carriers (including power storage) — J_4, d) production of final energy forms — J_5, e) CO_2 removal — J_6, $J_1,...,J_6 \in J$.

Each technology *j* is characterised (for the time period $t \in T$) by: the specific expenditures on product output — c_j, the specific consumption of resources and the outputs of products and harmful substances — a_{ij}, the constraints on minimum — b_j^{min} and maximum — b_j^{max} scales of utilisation. The yearly production of technologies (x_j) are the optimised variables of the model. Specific indices are taken constant during each calculated period *t*.

Due to the very long forecasting period the GEM-10R model was constructed as quasi-dynamic. The regional energy structure and the interregional fuel/energy exchange are optimised sequentially for the years 2025, 2050, 2075 and 2100. The obtained information (the optimal solution) is transferred as initial data for the next 25-year period.

Then the technological structure of the world energy system for period t can be represented mathematically in the following form.

Determine the minimum of the objective function

$$Z = \sum_r \sum_j c_{rj} x_{rj} \tag{2.1}$$

subject to (2.2) – (2.12).

1. For each region $r \in R$ it is necessary to fulfill:
a) the constraints on
    — primary energy mining (production)

$$\sum_{j \in J_1} a_{rij} x_{rj} \le b_{ri}, \qquad i \in I_1, \tag{2.2}$$

    — final energy production

$$\sum_{j \in J_5} a_{rij} x_{rj} \ge b_{ri}, \qquad i \in I_3, \tag{2.3}$$

    — electric power consumption modes (on power in typical points of electric load curves)

$$\sum_{j \in J_3} a_{rij} x_{rj} - \sum_{q \in J} a_{riq} x_{rq} \ge b_{ri}, \quad i \in I_4, \quad q \ne j, \tag{2.4}$$

    — emissions of harmful substances and CO_2 into the environment of a region

$$\sum_{q \in J} a_{riq} x_{rq} - \sum_{j \in J_6} a_{rij} x_{rj} \le b_{ri}, \quad i \in I_5 \quad q \ne j; \tag{2.5}$$

b) the balance equations for production/consumption of
    — primary energy

$$\sum_{j \in J_1} a_{rij} x_{rj} + \sum_{j \in J_2} a_{rij} x_{rj} - \sum_{j \in J_3} a_{rij} x_{rj} - \sum_{j \in J_4 J_5} a_{rij} x_{rj} = 0, \quad i \in I_1, \tag{2.6}$$

    — secondary energy carriers

$$\sum_{j\in J_4} a_{rij} x_{rj} + \sum_{j\in J_2} a_{rij} x_{rj} - \sum_{j\in J_3} a_{rij} x_{rj} - \sum_{q\in J} a_{riq} x_{rq} = 0, \quad i \in I_2,$$

$$q \neq j, \tag{2.7}$$

— electric energy and capacity

$$\sum_{j\in J_4} a_{rij} x_{rj} + \sum_{r\in R} \sum_{j\in J_2} a_{rij} x_{rj} - \sum_{p\in R} \sum_{j\in J_3} a_{pij} x_{pj} - \sum_{q\in J} a_{riq} x_{rq} = 0,$$

$$p \neq r, \ p \in R, \ i \in I_3, \ i \in I_4, \ q \neq j; \tag{2.8}$$

b) the constraints on the variables

$$b_{rj}^{\min} \leq x_{rj} \leq b_{rj}^{\max}. \tag{2.9}$$

2. For the world as a whole it is necessary to fulfill
   a) the global constraint on CO_2

$$\sum_{r\in R} \sum_{q\in J} a_{riq} x_{rq} - \sum_{r\in R} \sum_{j\in J_6} a_{rij} x_{rj} \leq b_{ri}, \quad i \in I_5, \quad q \neq j; \tag{2.10}$$

b) the balance equations of export/import of primary and secondary energy carriers and investments (description of the world markets of fuel and investments)

$$\sum_{r\in R} \sum_{j\in J_2} a_{rij} x_{rj} - \sum_{p\in R} \sum_{j\in J_3} x_{pj} = 0, \quad i \in I_1, I_2, I_6, \ p \neq r, \tag{2.11}$$

where

$$x_{rj} \geq 0, \qquad j \in J. \tag{2.12}$$

The model envisages two methods for representing export/import relations among the regions:
1) through the world market that is entered by each region independently;
2) by specifying direct ties among the regions (where they are real).

For electric power and gas transportation by pipelines the export/import ties are assigned directly in pairs, but only between the regions where they are actually possible.

Representation of the export/import ties via the world market essentially decreases the model dimension.

2.3.3. Computing system GEM-10R

To effectively apply the global energy model a special software package was worked out that was combined into the computing system GEM-10R.

The computing system consists of: 1) databases; 2) a control environment (a system manager); 3) a solver of linear programming problems; 4) a base of computation results.

The following databases were created:
— on energy resources;
— on demands for final energy forms and services;
— on environmental constraints;
— on financial and other constraints;
— a technological database (including description of the existing technological structure of each node).

The system manager consists of blocks: control of databases, preparation of initial data for optimisation, control of the solver, processing of computation results, preparation of tables and plots.

A system of electronic tables Quattro Pro is a key element of the user interface. It provides work with the bases of input data, preparation of data for the solver and representation of the computation results in tabular and graphic forms. Hence, the users can work in a consistent environment, quickly and qualitatively prepare output documents and take full advantage of all the resources and capabilities of the Quattro Pro package. The interface language is English.

The original technique for identification of objects that is applied in GEM-10R and the principle of "visual programming" that is realised by the interface resources allow easy modification of input data and reconstruction of the model. In particular, it is possible to add new regions, resources, products and objects, change the time horizon of the studies and steps, introduce additional constraints (environmental, social, etc.) in the form of equalities and inequalities, in particular on variables. There are possibilities for preparation and storage of the scenarios of external conditions and calculation results.

The model adjustment is considerably simplified owing to service programs, allowing a visual detection of the majority of errors even at the generation stage of the optimiser's input file.

The user can easily and quickly change the form and structure of input and output tables. The graphics was created by Quattro Pro. There are possibilities for export of initial data and calculation results to the tabular processor Excel.

In the model there are no rigid constraints on the length of the studied period, the number of considered regions or the number of optimised technologies, thanks to application of an effective opimiser for solving linear programming problems. Real capabilities of the model are practically fully determined by the computer resources applied. The time of computation and processing of the results is not critical (less than 10 min. on a Pentium-133 with 16 MB RAM for problems of rather high dimensionality — 4 thousand variables and 2 thousand equations).

The model was realised on a PC by the mathematicians I.Ya.Kavelin (the first version) and V.N.Tyrtyshny (the second version as the software package GEM-10R with an interface).

Chapter 3

ENERGY DEMAND

3.1. Peculiarities of energy demand forecasting

Energy demand forecasting is an indispensable and a very important component of studies on the long-term energy development at all territorial levels, i.e. national, regional, global. Forecasts are made on the base of different approaches and methods in all energy studies, starting from the works of the Club of Rome [71] and IIASA [2].

As was already noted, during the many-year studies the authors of this book generated several forecasts of energy demand for 10 separated world regions. The forecasts were made for four final energy forms (electrical, thermal, mechanical and chemical) for the years 2025, 2050, 2075 and 2100. The forecasting technique was gradually improved. Recently account has been taken of conditions and requirements of transition of the world community to sustainable development. The tendency to minimise energy demand is one of the key requirements.

It is difficult to quantify a minimum energy demand level, ensuring sustainable development, since it depends on historical traditions, lifestyle, really achievable technological progress, economic development of the regions and other factors. Therefore, use was made of expert judgements, comparisons and analogies, iterative calculations, etc. for its determination. Satisfactory estimates of the minimum necessary or sufficient energy demand can be obtained solely through laborious and long studies with an accumulation of experience. As the book was being written, the authors were at the initial stage of such studies. Thus, the below estimates of minimum energy demand are not final and will require adjustments at a later time.

The problem of determining the minimum necessary energy demand can not be approached as new. The book [5] was apparently a first work in this direction. The approach (a variety of the straightforward methods of calculations for individual processes) applied there is extensively used in the described studies. Specific energy demand estimates obtained in [5], however, seem to be very low and unlikely in the 21st century. Therefore, the authors of this book performed their own estimation of minimum energy demand for a developing region for the time span 2020–2030 on the base of a similar approach.

The approach applied by the authors to determine the minimum energy demand and the first results are presented below in Sections 3.2 and 3.3. Two variants (high and low) of energy demand forecast for the 10 world regions which were used directly for calculations by using the GEM-10R model are described in Section 3.4. The low forecast variant may be treated as minimal, however it was made by another method. Some results obtained in Sections 3.2 and 3.3 are compared to the forecast. More detailed comparison of the results obtained by different methods to make more reliable forecasts is planned in the future.

Section 3.2 presents a method of thermodynamic limits to determine theoretically minimal (and then actually achievable) energy consumption for output of energy-intensive products such as metal, chemicals, construction materials, etc. It allows an extension of the straightforward methods that are applied, as a rule, for short-and medium-term forecasts to the long-term forecasting of energy consumption in the industry. Its application to the main energy-intensive products (steel, aluminium, cement, etc.) is expected to result in forecasting of up to 70–80% of energy consumption in the industry with division into final energy forms (electrical, thermal, mechanical and chemical).

Section 3.3 describes application of the mentioned approach [5] for determination of the specific (per capita) energy consumption in developing countries. However, preconditions for the technological progress and provision of energy services for population are taken to be more realistic as compared to [5]. The general calculation scheme is also similar to the MEDEE models and the approach used there [72], but it is not very detailed because of the long-term forecasting horizon.

Two variants of the energy consumption forecast are described in Section 3.4 which were applied then to calculate characteristics of the world energy system by the GEM-10R model (Scenarios 1–6 – high energy consumption, 7 and 8 – low energy consumption, see Chapter 6). Energy consumption was determined based on changes in economic growth rates as a function of the achieved economic development level and decrease of the energy-GDP ratio owing to technological progress in energy production and consumption sectors. The variant of high energy consumption corresponds to maintenance of basically the existing trends in the economy and energy development of the regions (such scenarios are often called "business-as-usual"). The variant of low consumption correlates with a tendency to transition to sustainable development (e.g. financial and technical assistance of the developed to developing countries, intensive introduction of energy conservation technologies).

After the corresponding distribution of expenditures of the world regions on energy development was calculated by using GEM-10R, an additional study was performed on the macroeconomic model (Section 9.4) of the impact of energy price rise on its consumption. It showed that two chosen variants reflected reasonably well the possible (probable) changes in energy consumption depending on energy development conditions during the 21st century.

3.2. The method of thermodynamic limits

The method described below is a variety of normative forecasting methods (or the straightforward methods). Such methods are applied, as a rule, for the short-term (up to 5 years) or medium-term (from 5 to 10 years) forecasting. In this case the task was to modify the normative method, make it suitable for energy consumption forecasting for a long-term perspective.

The method is based on the thermodynamic analysis of production processes and the notions that were worked out to study energy efficiency and estimate energy conservation potential in different branches of the economy [65, 66]. It takes into consideration the technological progress in energy-intensive branches, decrease of energy consumption owing to improvement of existing processes and what is most important, analyses introduction of new technologies (including the most exotic ones from the current standpoint). This is essential in energy consumption forecasting for the future. As a result the minimum possible, thermodynamically conditioned energy consumption for output of the most important products is determined. It can not be lower even in the remote future at the highest level of engineering and technological development.

Any process to be accomplished requires an input of a certain amount of energy (electrical, thermal, energy of fuel from direct use) with provision of its appropriate structure. For forecasting it is essential to know first of all the total energy consumption, within which a fraction of every form of energy is determined.

An important characteristic of the described method is that besides the indicated traditional energy forms the total energy demand is forecasted considering the chemical energy of raw material [73]. Specialists of different branches of the industry (chemistry, metallurgy, etc.) run into manifestation of chemical energy of non-energy stores. Metallurgists know that the processes, whose raw material contains a large fraction of metal sulphides, liberate a great deal of energy (heat). It is precisely this energy that underlies autogenous processes of the ferrous and non-ferrous metallurgy and other branches. Increase in the level of using chemical energy of raw material is one of the most significant directions of the technological progress in industry in terms of decrease of both consumption of natural energy resources and releases of harmful substances into the environment. The notion of *chemical energy of a substance* plays an important role in the thermodynamic analysis of technical systems based on complex physicochemical transformations.

Numerical values of the parameters for calculations and forecasts of specific consumption of energy resources are determined by the thermodynamic effectiveness of real production processes. They vary within very wide ranges even for similar technologies in different countries. At the same time every technology (blast furnace process, steel-making processes, electrolytic production of aluminium, etc.) has a strictly definite limit of thermodynamic effectiveness independently of the country in which the enterprise operates.

3.2.1. Determination of limiting and real thermodynamic indices of production processes

Process efficiency is known to be the most suitable and comprehensive index of thermodynamic effectiveness. However, it can be determined solely for the processes to which the energy is supplied and whose useful product is also some energy form (or may be expressed in energy units). For numerous energy-consuming processes with production of metals, articles manufactured from them, industrial materials, chemicals, products, services providing comfort, etc. the efficiency can not be calculated in the commonly accepted sense.

This problem can be solved by substituting the *useful energy* in the numerator of the efficiency expression by the magnitude of *limiting (theoretically achievable minimum) consumption of energy or exergy* I_{min}^{*} for process accomplishment or output of some product. This magnitude is determined from the energy balance for an ideal analogue of the real process with $\eta_{en}^{id} = 1; \eta_{ex}^{id} = 1$ (the maximum degree of idealisation). Thus, the real energy efficiency of any production process is equal to the ratio of theoretical energy consumption to the factual one:

$$\eta_{en}^{real} = I_{min}^{*} / I_{en}^{real}.$$
(3.1)

Hence, it follows that the absolute energy efficiency of a real process (not equal to zero) can be determined only when the energy consumption of the ideal analogue is not equal to zero, i.e. at $I_{min}^{*} \neq 0$.

The ideal process is the highest degree of idealisation and corresponds (if the analogy is drawn) to the Carnot cycle in the power industry. However, the maximum degree of idealisation does not allow a deep analysis of real processes. Therefore, the *idealised analogues* of real processes (similar to the Diesel and Rankine cycles, etc. in the power industry) to be worked out must be closer to real conditions. They should include primarily two most important factors that influence energy consumption: the raw material used for product output and the technology applied.

The energy balance of the idealised analogue forms a base for determination of *the minimum necessary energy consumption* I_{min} for each studied process. The energy efficiency for the idealised analogue can be written as well:

$$\eta_{en}^{id} = I_{min}^{*} / I_{min}.$$
(3.2)

Comparison of the efficiency of real processes and idealised analogues via the relative efficiency that is extensively used in the energy sector

$$\eta_{en}^{rel} = \eta_{en}^{real} / \eta_{en}^{id},$$
(3.3)

makes it possible to estimate a degree of the studied process perfection. The higher is the value of η_{en}^{rel}, the more perfect is the process in the energy sense and the more difficult it is to find ways for its effectiveness improvement. The relative efficiency may be used to compare different production processes and to forecast a rational structure of technologies of product output.

Dependence on time of the efficiency of any real process in the course of its practical application and improvement is an ascending curve that asymptotically approaches some limit, i.e. the efficiency of the process, accomplished in idealised conditions — η_{en}^{id}. Correspondingly, energy consumption of the process for output of product unit tends with time to some practically unachievable magnitude — energy consumption of this idealised process — *the minimum necessary energy consumption.*

Note that the attained effectiveness of the same technology can differ in different countries, but they will have the same I_{min} and the idealised efficiency. If the analysis shows that further improvement of the technology gives rise to essential technical problems and financial expenditures, but leads only to negligible efficiency increase, this technology "has exhausted itself" and it should be replaced by the new more appropriate one.

The new technology can differ radically from the previous one. Therefore, its own idealised analogue should be worked out, its energy efficiency and specific energy consumption, whose values will be a reference point to perfect this new process, should be determined.

In order to use the information on variation of the thermodynamic effectiveness of technologies for energy consumption forecasting, the efficiency values must be extrapolated to the desirable forecasting horizon based on retrospective data, with the potential operation time of each technology being taken into consideration. Forecast of both the absolute and the relative efficiencies can be employed in this case. The latter index is preferable, as far as the ideal analogue is not always suitable for this purpose, as was already indicated.

Thus, variation of the energy efficiency during the production process life serves as information for forecasting its energy characteristics. Based on the minimum necessary energy consumption for output of the considered product and the forecast of technology efficiency at time interval t, the total specific energy consumption is determined:

$$I_t = I_{min} / \eta_{en_t}^{rel} . \qquad (3.4)$$

The structure of energy carriers for the given process, i.e. the fraction of electrical energy E, chemical energy of fuel I_f, chemical energy of raw material I_m, thermal energy I_q, is determined from studies of the real energy balances and the balances of the idealised analogue of a process.

Hereafter these values can be applied to forecast energy consumption independently of whether the forecast is made on the total energy consumption, when the volume of gross output of the considered product is given, or on the per-capita energy consumption. The only requirement is that the forecast should be in physical rather than in cost units.

The authors called the devised method the *method of thermodynamic limits.*

In order to determine formation principles of the idealised analogues for the whole variety of energy-consuming processes in different spheres of the economy they were classified in the following way [66]:

1. *Energy production processes* underlying technical systems that are aimed at output of different energy products.

2. *Physicochemical processes of production of materials, chemicals and other substances.* This group includes all the processes which form a base of chemical and

metallurgical productions, for example, smelting of metals, their alloys, production of fertilisers, plastic, binders, dyes, insulating materials and other products.

3. *Processes of production of articles, work accomplishment.* In technical systems designed for production of articles from the given material, the worth of the product can not be expressed in energy units. For example, machine tools, cars, clothes, shoes can not be characterised by energy. The thermodynamic effectiveness of such processes should be estimated by using the values of minimum necessary energy consumption for their implementation, as in the previous case.

4. *Processes for providing normal conditions of life and work for people.* The considered group comprises heating, ventilation, air cooling, lighting of different kinds of buildings, refrigerating and freezing of perishable products, etc. The indicated processes are characterised by the fact that their implementation results in some useful consumer effect.

5. *Movement of people and freight in space* includes diversity of transportation modes by all transport means: railway, water, air, automobile.

Each group of energy-consuming processes is characterised by its particular principles of generating idealised analogues. They have been worked out to the greatest extent for the second group of processes (to be considered below), which allows practical forecasts to be made.

3.2.2. An algorithm of energy consumption forecasting on the base of limiting thermodynamic characteristics of technologies

We consider the forecasting scheme of energy demands for manufacture of n energy-intensive products $i = 1, ..., n$ without separation of specific branches of industry. Each product can be manufactured by m technologies $j = 1, ..., m$ (the index i at m is omitted for simplification). Assume that the total energy consumption for manufacturing n products (from the second group of processes) will make up a considerable portion of total energy consumption (in industry or the economy as a whole).

Figure 3.1 presents a general block-diagram of the forecasting process. It begins with determination of the list of n energy-intensive products, for which the calculations will be performed (Block 1). The mix of these products is selected considering their potential fraction in total energy consumption, available information, admissible labour input (and terms) for making forecasts and other factors.

Then each product i is analysed in turn. On the one hand, the specific (per capita) or the total demand for it in the considered country or region (Block 2) is determined and on the other hand, each particular technology $j = 1,..., m$ to manufacture this product is analysed and the required calculations are performed (Blocks 3–5).

For each technology of the considered mix an idealised analogue is elaborated, the minimum necessary energy consumption I_{min} and the idealised energy efficiency η_{en}^{id} are determined. The calculations are made to establish upward tendencies of the factual efficiency of the technologies considered and a forecast is made for the future (η_{en_t} — at the forecasted level).

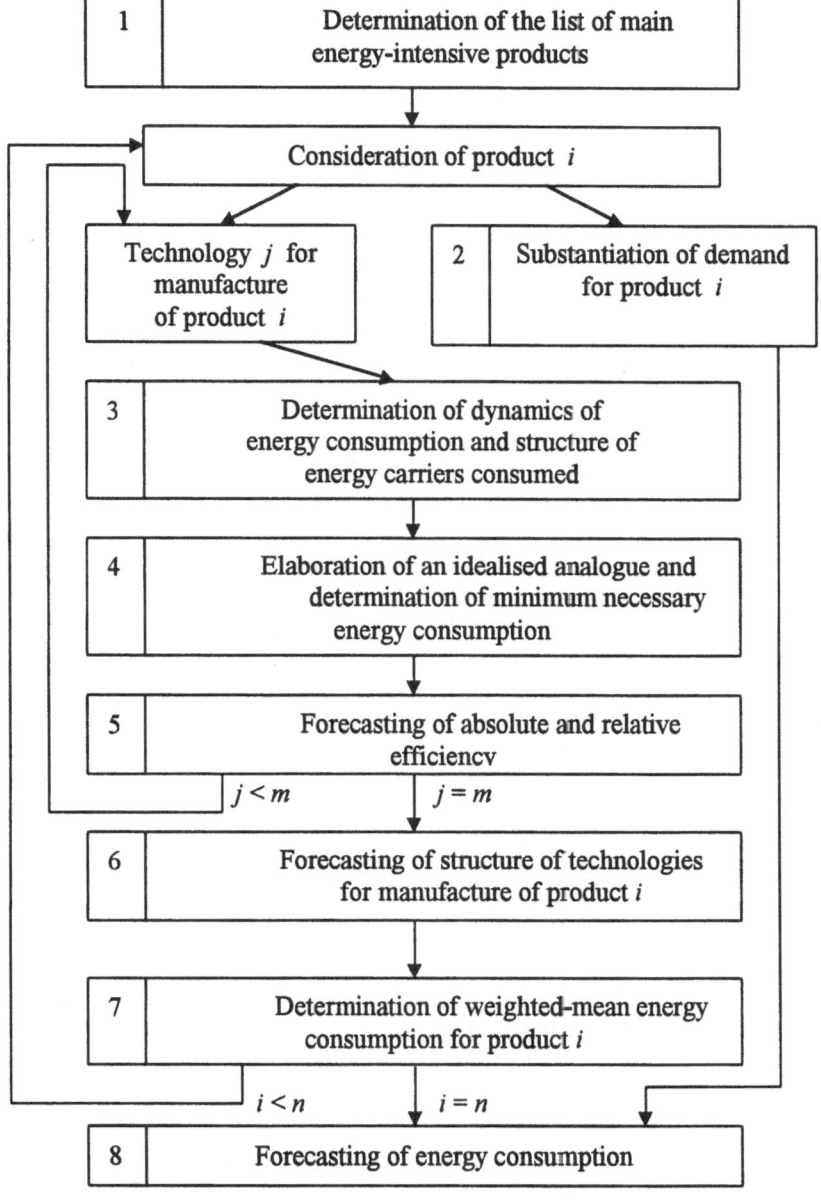

Figure 3.1. Block-diagram of energy consumption forecasting.

These indices are used to determine relative energy efficiency of the technology:

$$\eta_{en_t}^{rel} = \eta_{en_t} / \eta_{en}^{id} \tag{3.5}$$

and specific energy consumption for the forecast:

$$I_t = I_{min} / \eta_{en_t}^{rel}. \tag{3.6}$$

While the limiting energy consumption for productions can be determined from the ideal analogue, the specific consumption can also be expressed via the absolute efficiency, i.e.

$$I_t = I_{min}^* / \eta_{en_t}. \tag{3.7}$$

As was noted, such processes cover technologies of ferrous and non-ferrous metallurgy, chemical and petrochemical industries. Any described calculation method is suitable for them, i.e. a forecast is made either on the base of the absolute or the relative efficiency.

Along with the total energy consumption for each technology the blocks contain its distribution among individual energy forms: electrical, thermal, chemical energy of fuel of direct use, and raw material.

The next step of forecasting is to determine an optimal structure of the technology for manufacturing product i (Block 6). This problem is solved in terms of the minimum discounted costs by applying additional criteria such as the minimum of energy consumption, capital investments, labour inputs, releases into environment, and the maximum of equipment capacity, etc. When determining an optimum structure of technologies for ferrous metallurgy, an additional condition can be laid down on the maximum possible use of scrap metal, because this is one of the most significant energy conservation measures.

This problem can be solved by the methodical approach based on forecasting the process of replacement of existing technologies by new ones. The model constructed in [74] represents further improvement of the Fischer–Pry model [24] intended for forecasting technological changes on the base of the historical data on replacement of one technology by another.

After the quantity of product i manufactured by different technologies is determined, the weighted-mean energy consumption per unit of product i is calculated (Block 7). It allows then the overall energy consumption for its manufacture with regard to the demand, forecasted for product i in Block 2, to be determined (Block 8).

Operations described for Blocks 2–7 are executed for all n energy-intensive products. Then the total energy consumption for the considered region and time horizon is calculated in Block 8. In so doing its division into individual energy forms is also calculated.

Numerous investigations show that the accounting of output of basic energy-intensive branches of industry as well as energy services rendered to the population makes it possible to describe no less than 70% of energy consumption in any region. The rest of energy consumption can be taken into account by the coefficient μ. Hence, the energy demand for the forecasting horizon t can be determined by the devised technique as follows:

$$\overline{I}_t = (1 + \mu_t) \left[\sum_{i=1}^{n} \sum_{j=1}^{m} \alpha_{ij_t} \overline{M}_{it} I_{\min_{ij}} \eta_{en_{ij}}^{rel} + \overline{P}_t \sum_{i=1}^{n} k_i \sum_{j=1}^{m} \alpha_{ij_t} I_{\min_{ij}} \eta_{en_{ij}}^{rel} \right]. \quad (3.8)$$

Here i — an index of product, service, useful effect; j — an index of technology for manufacture of product i; t — the forecasting horizon; M — the volume of product output; α_{ij} — a fraction of product i manufactured by technology j; P — the population number, k_i — demand for product i per capita. The values with the bar above are the forecasted values.

The first component takes into account energy consumption for product output that is set by the total volumes. The second considers energy consumption calculated per capita (for the life support system: heating, ventilation, air cooling, hot water supply, cooking, lighting, etc.; for the system of all transport means; different kinds of services).

Due to uncertainty of initial information the value I_t lies in a range which widens with the increasing depth of forecasting.

For branches of the industry this forecast can be made for both total and specific consumption. However, the other spheres are characterised, as a rule, by the per-capita energy consumption. Therefore, for the purposes of uniformity it is possible to forecast specific indices of output of different products per capita and then to pass to total demands through the forecast of population number. Thus, the devised algorithm may be applied for energy consumption forecasting on the base of historically formed trends and also in the case when they are violated. Such a problem can arise when the idea of *sustainable development of the world* is implemented. All these factors can lead to transformation of the problem of energy consumption forecasting. Whereas it is solved routinely by the principle *from the past to the future*, the necessity may emerge to apply the inverse principle: *from the future to the present*. In such a statement the problem is more successfully solved by the specific indices (demographic, economic, energy).

The final stage of forecasting, as was mentioned, is to distribute the total energy demand among its individual forms: chemical energy of raw material, fuel, electrical and thermal energy in accordance with the structure of energy carriers used in basic energy-intensive technologies.

3.2.3. Production processes of substance with a given chemical composition

Consider the second group of processes by our classification. The ideal analogue of producing some substance with a given chemical composition is a reversible chemical reaction with $\eta_{en} = 1$ and $\eta_{ex} = 1$. Such a property is characteristic of the devaluation reaction of this substance, when its chemical energy i_x and exergy e_x are determined [65, 73]. Thus, the limiting minimum energy consumption I_{min} for production of substance k is determined by the magnitude of its chemical energy. This consumption is absolutely minimal for production of this substance irrespective of the used raw material that is extracted from the environment and the technology of its processing. Thus, we can write

$$I^{*}_{min_k} = i_{x_k},$$

i.e. the corresponding values of chemical energy can be taken as the limiting energy consumption for processes of the second category. For example, Table 3.1 presents the values of *i* for some industrial products: metals and chemicals.

Table 3.1. Limiting (theoretical) energy consumption for production of some metals, inorganic and organic substances, GJ/t

Product	Chemical formula	I_x, GJ/t	Product	Chemical formula	I_x, GJ/t
Metals					
Iron	Fe+~4%C	9.510	Soda ash	Na_2CO_3	1.022
Carbon steel	Fe+to 1.7%C	8.954	Caustic soda	NaOH	3.449
Aluminium	Al	35.360	Sulfuric acid	H_2SO_4	0.971
Lead	Pb	1.544	Sulfur	S	19.464
Chromium	Cr	12.656	Chlorine	Cl	0.705
Copper	Cu	3.541	*Organic chemicals*		
Manganese	Mn	9.501	Ethanol	C_2H_5OH	29.744
Nickel	Ni	6.217	Benzene	C_6H_6	41.904
Silver	Ag	0.788	Urea	$CO(NH_2)_2$	10.542
Zinc	Zn	7.006	Calcium carbide	CaC_2	24.680
Tin	Sn	5.004	Methanol	CH_3OH	22.720
Titanium	Ti	19.729	Ethylene	C_2H_4	50.404
Inorganic chemicals			Acetic acid	$C_2H_4O_2$	14.588
Ammonia	NH_3	20.509	Acetylene	C_2H_2	49.996
Hydrogen	H_2	143.000	Isoprene	C_5H_8	46.486

The following general principles are accepted in working out the idealised analogues of production processes involving chemical transformations.

The process that is described by one (or several) basic irreversible reactions, is taken as such an analogue, as in the real manufacture of the considered product. In such an idealised process chemically pure compounds are used and a chemically pure product is obtained.

There is no loss of materials and the reactions proceed to completion with stoichiometric amounts of reactants.

The temperature of input and output substances is taken equal to the ambient temperature. It is obvious that such a process proceeds with minimum energy consumption.

Combining the idealised analogues of individual processes in accordance with a selected flow chart results in analogues of different real productions.

The essence of the proposed method for estimation of energy consumption for the given classification group is shown on an example of aluminium production by the Hall–Heroult method that is common for all countries. Aluminium's steady improvement in terms of design and technology has turned it from a rare metal whose value was higher than silver to one of the most widespread and cheapest of metals, ranking next to steel among structural metals. Specific electricity consumption for alumina electrolysis has decreased therewith more than thrice.

Aluminium is produced by electrolytic decomposition of alumina melt in cryolite. Schematically the electrolysis is reduced to extraction of metallic aluminium on the cathode and combustion of the carbon anode owing to separation of oxygen on it.

According to the energy balance, aluminium is produced by consumption of electrical energy E, chemical energy of fuel in the form of carbon (anode mass or roasted anodes) I_f, and chemical energy of raw material (alumina) I_m.

The idealised analogue of electrolysis is described by the same reaction as the real process:

$$E + Al_2O_3 + 2C = 2Al + CO + CO_2.$$

The material and complete energy balances of this analogue are presented in Table 3.2.

Table 3.2. Material and complete energy balances of the idealised analogue of aluminium electrolysis (electrolysis with oxidised anodes), 1t Al

Item	Mass, t	Energy, GJ/t	Item	Mass, t	Energy, GJ/t
	Input			Output	
Al_2O_3	1.889	4.292	Al	1.000	35.360
C	0.443	14.533	CO	0.518	5.241
E	–	21.776	CO_2	0.815	0.000
Total	2.332	40.601	Total	2.333	40.601

From the complete energy balance the efficiency of the idealised analogue was $\eta_{en}^{idl} = 0.871$. The minimum necessary energy consumption for production of 1 t Al is equal to 40.601 GJ, of which 53.6% is electrical energy, 10.6% — chemical energy of fuel, 35.8% — chemical energy of raw material, i.e. the energy consumption structure can be described by the following relation:

$$E : I_f : I_m : I_q = 0.536:0.106:0.358:0.0.$$

Figure 3.2 illustrates changes in specific consumption of all kinds of energy carriers for the process (a) and the energy efficiency (b) over the time interval of process commercialisation in world practice. The use of electrical energy and carbon materials is seen to decrease to the greatest extent. Factual total consumption asymptotically approaches a value of minimum necessary consumption. The main options for decreasing energy consumption for this technology have been practically exhausted. These are growth of the unit capacity of electrolysers, automatic control of the process by checking alumina concentration in electrolyte, decrease of voltage drop in the bath, improvement of electrode quality, uninterrupted supply of alumina to baths, perfection of buses, and use of lithium salts as an additive to electrolyte. Still there are reserves for decrease, however its rates steadily slow down and further decrease of energy consumption will cause great technical problems and financial expenditures.

Thus, in the electrolysis process with consumable anodes the potentials for enhancing the efficiency have been practically exhausted. Besides, recently the carbon materials required for production of anode mass and roasted anodes (petroleum and pitch coke, anthracite, pitch, etc.) have become sharply deficient and requirements for their quality have also become more strict. The specialists show, therefore, an increasing interest to designing the electrolysis process with non-consumable (inert) anodes.

There are no technical and economic indices and in particular, no specific energy consumption of such installations in publications as yet. The process can be analysed at a level of the idealised analogue and conclusions can be made on its possible thermodynamic effectiveness, structure of energy carriers used and hence, its prospects.

The idealised analogue of the aluminium production process with inert anodes is described by the reaction:

$$Al_2O_3 = 2Al + 1,5O_2 .$$

The material and complete energy balances of this process are presented in Table 3.3 and its energy efficiency is $\eta_{en}^{idl} = 1.0$.

Thus, the calculations performed for idealised analogues show that the specific electricity consumption for production of 1 t Al with inert anodes is 1.43 times as high as with oxidisable ones. Total energy consumption in this case, however, is lower and the efficiency is higher than with consumable anodes. When implementing this process there will be no need for large amounts of high-quality and deficient carbon materials to produce anode mass and anodes. Besides, the studies reveal that when processing unoxidisable anodes in electrolysers, anode effects are excluded, liberation of fluoric gases decreases, and regulation conditions improve.

a)

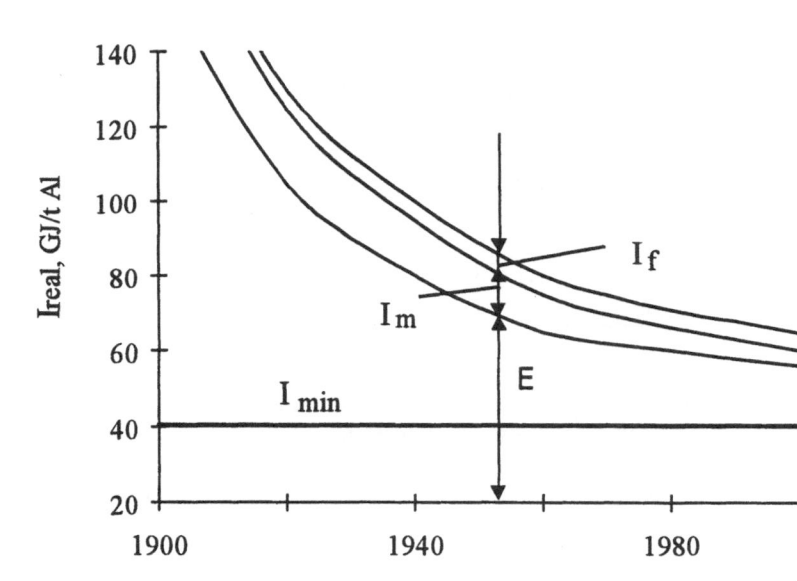

b)

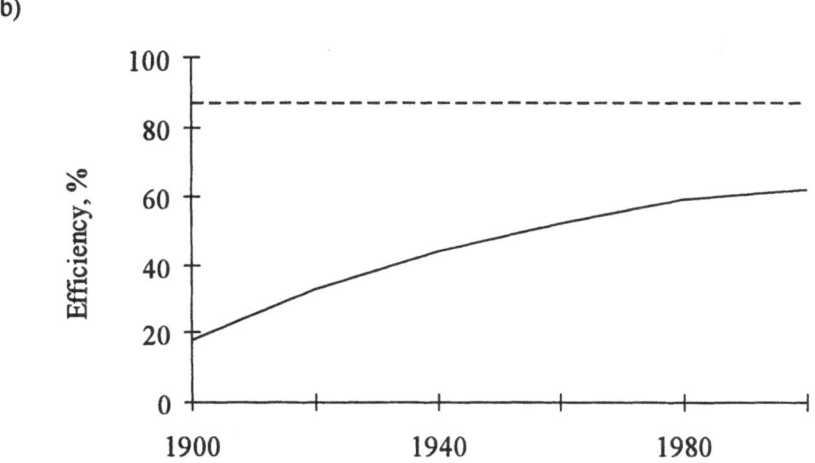

Figure 3.2. Dynamics of energy characteristics of aluminium production process by the Hall–Erow method. a — total consumption of energy and its individual forms; b — relative energy efficiency (efficiency of the idealised analogue is represented by the dashed line).

The use of anodes made of alloys Al, Sn and Be results in a four-fivefold reduction of the interanode distances and hence, a 20–25% decrease of electricity consumption. As practice with the process increases, it will be technologically and structurally improved, leading to a decrease in energy consumption (as was the case with the current method) (see Figure 3.2a).

Table 3.3. Material and complete energy balances of the idealised analogue of aluminium electrolysis process (electrolysis with inert anodes), 1t Al

Item	Mass, t	Energy, GJ/t	Item	Mass, t	Energy, GJ/t
	Input			Output	
Al_2O_3	1.889	4.292	Al	1.000	35.360
E	–	31.068	O_2	0.889	0.008
Total	1.889	35.360	Total	1.889	35.368

The indicated factors show with sufficient evidence the necessity of a detailed analysis of the electrolysis process with inert anodes. The final choice between them, however, can be made by comparison of these processes when all expenditures, financial, material and energy as well as the related productions (anode production, for example), are taken into consideration.

From the above it follows that energy demand forecasting directly depends on the forecast of technological progress in the considered branch. In making scientific and technological forecasts, the possibility of designing new machinery is determined, changes in the technological level of production and product quality are studied and planned, and the appropriate expenditures and technical and economic indices for new machinery are calculated.

Table 3.4 presents limiting energy characteristics of the production process of some metals by different technologies.

In the future, even in 50–100 years, mankind will assumedly utilise the same materials as at the present time: metals (steel, aluminium, copper, nickel, zinc, titanium), major construction materials (cement, brick, glass), chemical substances and chemical fertilisers. New synthetic materials will probably appear in addition. People will have similar, but better living conditions: comfort (heating, ventilation and air cooling), cooling/freezing of food products, all means of transport services, full-value nourishment and climate-appropriate clothes.

The basic technologies utilized in the output of the world's main product types are, as a rule, of great vitality. The blast-furnace process has been in operation for no less than 150 years and the technologies of a comparatively "young" structural material — aluminium — achieved 100 years of age in 1986. Many technologies being used now will exist in the future. However, some of them will be replaced by new ones, either because of their low efficiency or due to the changing quality and type of raw material processed.

Table 3.4. Energy indices of idealised analogues of production of some metals

Product, technology	Minimum necessary energy consumption, GJ/t	η_{en}^{idl}, %
Steel smelting by technology:		
Iron + steel smelting	14.423	59.2
Electric smelting of scrap	2.692	88.1
Direct iron reduction	3.077	77.1
Aluminium (*from alumina*)	40.601	87.1
Copper (*pyrometallurgical processes*)	28.010	12.6
Lead (*blast smelting*)	6.002	25.7
Zinc		
Pyrometallurgical methods	23.063	30.3
Hydrometallurgical method	20.288	34.6
Titanium (*magnesium–thermal method*)	37.216	53.2
Silicon (*electric smelting*)	52.571	61.6

The scope of work to be done to adequately understand energy characteristics of comparative technologies is extremely large. Nonetheless, the detailed forecasting of energy development and selection of the structure of energy technologies are impossible without a qualitative forecast of energy consumption. Therefore, energy studies of branches of the economy should be a critical component of the work on energy consumption forecasting for a long-term time horizon.

3.3. Sustainable energy consumption in developing countries

The forecast in [5] was made based on the assumption that the economic development level of developing countries will gradually approach the level of industrially developed countries.

A straightforward method based on separation of the most energy-intensive goods and services, analysis of dynamics of their energy intensity and the forecast of demands for them was used in [5]. An attempt was made to determine the minimum necessary energy consumption for some conventional developing country. For this purpose the final energy consumption was divided into several sectors: residential, commercial, transport and industry.

Technical processes, consumer devices and equipment that consume the greatest portion of energy are considered in each sector, the lowest currently possible specific energy consumption being taken per unit of useful output.

Hot water supply, kitchen stoves, refrigerators, TV sets, washing machines and lighting are considered for residential buildings.

In commercial buildings division of energy consumption by device was not done, specific energy consumption was taken equal to 0.13 GJ/m^2, as in the most energy-effective building in Sweden in 1981, minus energy consumption for heating.

All main types of passenger and freight transport: household motor vehicle, inter-city buses, passenger railway, city public, passenger air, freight motor, freight railway and water transport are separated in the transport sector.

In heavy industry consideration is given to production of steel, cement, aluminium, paper, fertilizers, construction, agriculture.

Consumption of the indicated materials, numbers of instruments and transport units in operation are taken the same as in developed countries in the beginning — the middle of the 1970s. As a result the value of final (electrical energy and fuel for consumers) per-capita energy consumption obtained was equal to about 1 kW·year/capita per year (without energy consumption for space heating and air cooling). If non-commercial energy is taken into account, this value will be close to the actually achieved level. Hence, the calculations in [5] show that, at the maximum increase of the effectiveness of final energy use, the developing countries can reach the standard of living in developed countries without increase in specific energy consumption. Such a result, however, seems hardly probable and not adequate for real capabilities of developing countries.

As noted in Section 3.2, the straightforward method is convenient to forecast energy consumption for sustainable development. It allows the forecast to be made in accordance with demands placed on the economic and social growth of developing countries.

The calculation of final (electrical energy and fuel) per-capita energy consumption of developing countries by using the straightforward method is presented below. It aims at determination of the scale of final per-capita energy consumption in developing countries assuming that the standard of living of their population will be approximately the same as in developed countries in the early 1970s and the technologies and equipment used will be no worse than in developed countries in the early 1990s. According to the GDP forecast (see Section 3.4) the per capita income in developing countries will reach a level of developed countries of the early 1970s (US$ 6–7 thousand/capita per year) in 2025–2030. The calculation was performed by the same scheme, as in work [5].

3.3.1. Residential buildings

The least specific (per capita) area of residential buildings for developed countries was in Japan in the mid-1970s and made up 25 m^2/capita [21]. This value is assumed as minimum required for comfortable living. It is also supposed that for every four people there is on the average a TV-set, a refrigerator, a washing machine and a stove.

3.3.2. Heating and cooling

Most energy in residential buildings is consumed to provide favorable microclimate, i.e. for heating and cooling. Energy consumption for these needs depends greatly on climatic conditions which can be characterized by the number of heating and cooling degree-days (HDD and CDD, respectively) in a year.

The number of heating degree-days for a day is the sum of differences of the indoor temperature taken equal to 18.3^0 C [75] and the daily average outdoor temperature (when the difference is positive) for all the days in a year. The definition of cooling degree-days is similar.

Figures 3.3 and 3.4 based on data for the USA [75–79] show that energy consumed for cooling is quite comparable with energy consumed for heating. With the number of heating degree-days less than 2500, and cooling degree-days more than 2500, energy consumed for cooling becomes larger than for heating (see Figure 3.3).

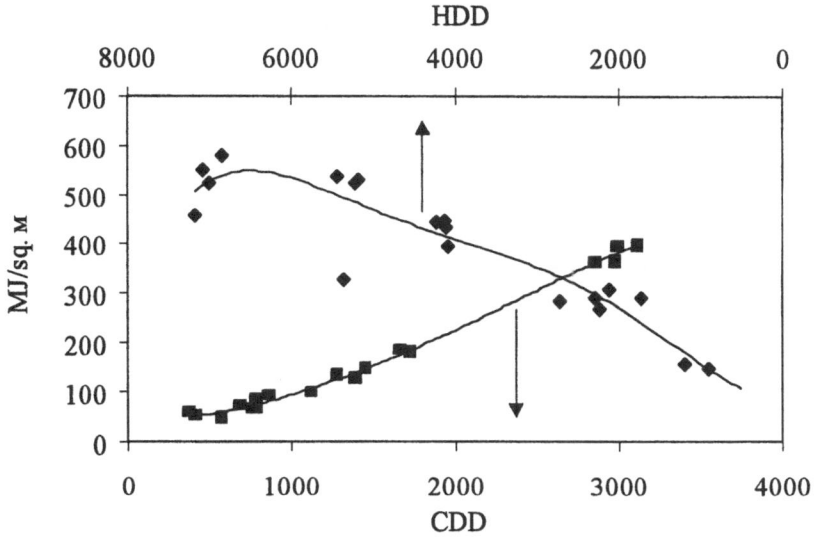

Figure 3.3. Relation between specific consumption of primary energy for heating and cooling of residential buildings and climatic characteristics.

Cold (more than 4000 heating degree-days) and hot (more than 2000 cooling degree-days) climates can be considered unfavorable in terms of energy consumption. It is seen from Figure 3.4. that among five climatic zones of the USA (Table 3.5) the least total energy consumption for heating and cooling is observed in zone 4 (HDD<4000 and CDD<2000). In the colder zones (1–3) energy consumption rises due to increase in energy consumed for heating and in the hottest zone (5) due to larger energy consumption for cooling. Here the total energy consumed for heating and cooling in the coldest and hottest climatic zones are practically equal.

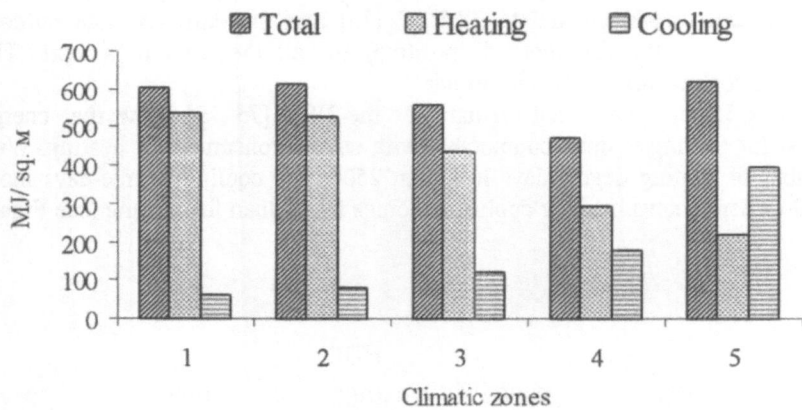

Figure 3.4. Primary energy consumed for heating and cooling of residential buildings.

Table 3.5. Climatic zones of the USA

Zone	HDD	CDD
1	> 7000	< 2000
2	5500–7000	< 2000
3	4000–5499	< 2000
4	< 4000	< 2000
5	< 4000	> 2000

Energy consumed for heating and cooling that was not taken into account in [5] has an essential effect on the total energy consumption and should be considered when forecasting energy demand of residential and commercial buildings.

Specific final energy consumption for heating and cooling of the residential buildings assumed in the calculation is averaged over climatic zone 5 with the number of heating degree-days less than 4000 and the number of cooling degree-days more than 2000 [75–78], namely: 177.3 MJ/m^2 for heating, 131.7 MJ/m^2 for cooling yearly.

3.3.3. Other consumers

When calculating energy consumption for hot water supply, lighting and domestic appliances the specific annual energy consumption was assumed somewhat lower than the average value for developed countries in the 1990s [75–79]:

Hot water supply, GJ/capita/yr   — 2.83
Lighting, GJ/capita/yr   — 1.08
TV-sets, GJ/unit/yr   — 1.45
Refrigerators, GJ/unit/yr   — 4.66
Stoves, GJ/unit/yr   — 4.57
Washing machines, GJ/unit/yr   — 3.97
Others, GJ/capita/yr   — 0.50.

Unlike [5] where specific energy consumption for the enumerated appliances was taken equal to the lowest possible, the given calculation assumes values meeting the real conditions. When analysing statistical data [75–79] the average values of the specific energy consumed by domestic appliances of the USA and some European countries were determined. Their slight decrease was taken for the calculation with allowance for possible enhancement of energy efficiency.

Table 3.6 presents specific energy consumption per unit of equipment, per unit of area and per capita in the residential sector. Thus, the total annual consumption of final energy makes up 15.79 GJ/capita, which is 4 times higher than indicated in [5], mostly due to large specific energy consumption for equipment as well as for heating and cooling.

Table 3.6. Annual energy consumption in residential buildings

Consumer	Energy consumption		
	Per unit of equipment, GJ/unit/yr	per capita, GJ/capita/yr	per unit of area, GJ/m^2
Heating		4.43	177.3
Cooling		3.29	131.7
Hot water supply		2.83	
Domestic appliances		5.24	
Refrigerators	4.66	1.17	
Lighting		1.08	
TV-sets	1.45	0.36	
Stoves	4.57	1.14	
Washing machines	3.97	0.99	
Others		0.50	
Total		15.79	

3.3.4. Commercial buildings

In the mid-1970s the minimum area of commercial buildings in developed countries was also in Japan and made up 8 m^2/capita, which is assumed as the minimum required area for developing countries.

   Similarly to residential buildings, energy consumption for heating and cooling
makes up a considerable fraction in energy consumption by commercial buildings.
Figure 3.5 presents the relation between the specific consumption of primary energy
and climatic characteristics. The conditions in the hottest climatic zones (4 and 5)
are most favorable, however, it can be supposed that at the hotter climate the total
energy consumption will increase due to rising energy consumption for cooling.

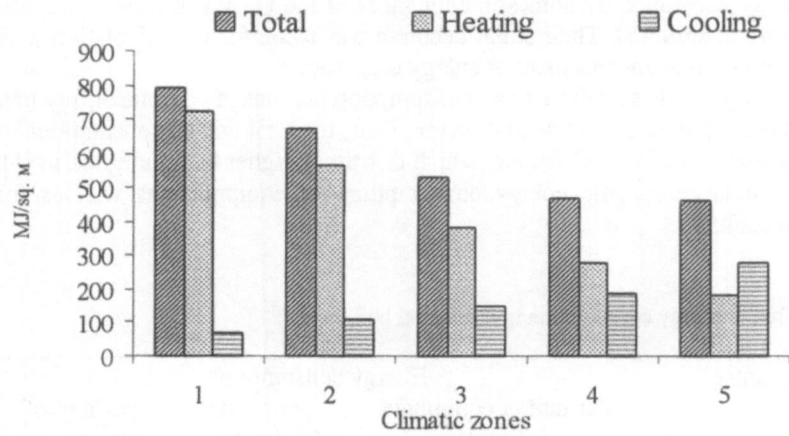

Figure 3.5. Primary energy consumption for heating and cooling of commercial buildings.

   Final energy consumption (MJ/m² per year) for heating, cooling and ventilation
was taken equal to the average value for climatic zone 5 with HDD<4000 and
CDD>2000 [78–82]:

heating	–	151.0
cooling	–	92.0
ventilation	–	51.1.

   Specific energy consumption for lighting and equipment per 1 m² of area was
assumed somewhat lower (supposing enhancement of energy efficiency) than the
average energy consumption in developed countries in the early 1990s according to
[78–82]; these data are presented in Table 3.7. Most of the energy consumed in
commercial buildings is due to lighting and office equipment. The total specific
energy consumption in public buildings totals 5.60 GJ/capita.

Table 3.7. Annual energy consumption in commercial buildings

Consumer	MJ/m^2	GJ/capita
Heating	151.0	1.20
Cooling	92.0	0.74
Ventilation	51.1	0.41
Hot water supply	5.7	0.05
Lighting	166.9	1.34
Cooking	11.4	0.09
Refrigerators	36.3	0.29
Office equipment	60.2	0.48
Others	124.9	1.00
Total	699.5	5.60

3.3.5. Transport

The following groups were distinguished:
— household motor vehicles;
— passenger transport;
— cargo transport.
It is supposed that the amount of household motor transport in developing countries will be determined by the ratio: 1 car per 8 people which again corresponds to Japan's level, the minimum of all developed countries. Fuel consumption and mileage are taken based on the example of Japan and France and make up 6.5 l (164 MJ) per 100 km and 10500 km per car yearly.

Intensity of passenger and cargo transport use was taken close to the magnitude of developed countries of the early 1970s according to [5, 83]; specific energy consumption for the enumerated kinds of transport was assumed somewhat lower than the level of developed countries in the early 1990s. The total annual energy consumption in the sector makes up 8.86 GJ/person (Tables 3.8–3.10).

Table 3.8. Energy consumption by household motor vehicles

Number of cars per family	0.5
Number of cars per person	0.125
Fuel consumption per car, 1/100 km	6.5
Fuel consumption per car, MJ/100 km	164
Annual mileage of a car, km	10500
Annual energy consumption, GJ/capita	2.16

Table 3.9. Energy consumption by passenger transport

Kind of transport	Mileage, pass.-km/capita	Energy intensity MJ/(pass.-km)	Energy consumption GJ/capita
Buses	1850	0.6	1.15
Public municipal transport:	520		
Electrical	312	0.41	0.1
Diesel	208	1.13	0.2
Railway transport:	3175		
Electrical	2223	0.2	0.4
Diesel	953	0.6	0.6
Air transport	345	2.5	0.9
Total			3.35

Table 3.10. Energy consumption by cargo transport

Kind of transport	Mileage, t-km/capita	Energy intensity, MJ/t-km	Energy consumption, GJ/capita
Motor transport	1495	0.8	1.20
Railway	814	0.7	0.57
Water transport			1.58
Total			3.35

3.3.6. Industry

The most energy intensive industries are ferrous and non-ferrous metallurgy, chemical industry, and production of construction materials. A forecast of consumption of products of these industries and specific energy consumption for their production makes it possible to determine the main share of energy consumption in the sector; energy consumption in the other sectors can be taken into account using a correction factor.

Dynamics of consuming energy intensive materials (steel, aluminium, cement) is similar to the dynamics of energy consumption. During the early phases of an industrialisation period, consumption of these materials grows fast due to the need for production infrastructure and due to the fast rise in industrial production scales. Further along in the industrial period, consumption growth rates of these materials stabilise and become proportionate to the economy growth rates. And, finally, in the post-industrial period, consumption growth rates of energy intensive materials decrease, their per capita consumption stabilises owing to the so-called dematerialisation of the economy, i.e. the industry fraction in the GDP of a country decreases because during this period the economy grows mainly at the expense of the service sector . These trends will be retained for developing countries as well.

Specific energy consumption for production of the above materials is supposed to be determined later using the methodology described in Section 3.2. Energy consumption forecast in the production sector has not been completed and for the present calculation was taken from [5] (Table 3.11).

Table 3.11. Annual consumption of final energy in industry

Material, sector	GJ/capita
Steel	3.32
Cement	1.90
Aluminium	1.17
Paper and paperboard	1.11
Fertilizer	1.14
Agriculture	1.42
Construction, mining	1.86
Others	8.75
Total	20.67

Thus, in 2025–2030 the total energy consumption for all sectors in developing countries will make up about 51 GJ/capita per year (Table 3.12), which exceeds the results obtained in [5] by 50%. Recall that this magnitude is a sum of the consumed electrical energy and end-use fuel. Having presented energy consumption in the form of a total of electrical, thermal, mechanical and chemical energy, the specific final energy consumption equal to 36–40 GJ/capita per year is obtained. As will be shown in Section 3.4, this magnitude is close to the energy consumption of the LA region in 2025 (high energy consumption) or in 2050 (low energy consumption) assumed for the calculations.

Table 3.12. Annual consumption of final energy in developing countries in 2025–2030

Sector	GJ/capita	%
Residential	15.79	31
Commercial	5.60	11
Transport	8.86	17
Industry	20.67	41
Total	50.92	100

3.4. Energy demand projection for the GEM-10R model

The need for final forms of energy (electrical, thermal, mechanical and chemical) for the world regions and calculated periods is initial information for the mathematical model GEM-10R. Energy consumption affects greatly competitiveness of energy technologies since it influences the depletion rate of cheap fossil fuel resources.

In the long-term future energy consumption is determined by the population number, level of economic development, energy efficiency and life style. Mathematical models of different complexity that connect energy consumption with the indicated (and some other) factors are applied to forecast energy consumption. Experience shows that energy consumption forecasts for a 20 year time span and longer can not be highly accurate. However if we consider the period of 100 years (as for instance, in the present work) the energy consumption forecast is more likely to be the hypothesis of possible, probable or desirable development paths. Therefore, in a number of cases it is expedient to apply a maximum, simple (in terms of information and mathematical methods used) procedure of forecasting.

3.4.1. Method

Two variants of consumption forecast are described below for four kinds of final energy in 10 world regions up to 2100. The forecast is based on extrapolation of the existing rates of economic growth and trends of change in the energy-GDP ratio in developed countries (decrease in energy intensity with a rate of about 1% yearly) and on the assumption that developing countries will repeat (to some extent) their course of development. In the "H" (high) variant of energy consumption the energy-GDP ratio of developing countries corresponds to the energy-GDP ratio of developed countries at the corresponding stage of economic development (at the same specific GDP) with a shift of 15 years ahead, which takes into account decline in energy intensity due to technological progress in the energy consumption, sector. The "L" (low) variant of energy consumption, which meets best the principles of sustainable development, supposes financial assistance of developed countries to developing ones and faster decrease in the energy-GDP ratio (a 35-year shift) provided by the extension of international cooperation in the sphere of science and technology.

The forecast was made, based on a small number of assumptions, and formalised. At the same time this procedure is rather qualitative and the results can hardly be sufficiently substantiated. Therefore we performed an informal expert analysis of current trends in consuming energy services by population and by sectors of the economy in developed and developing countries, as well as a long-term forecast of efficiency of energy technologies. Here consideration was given to the numerous forecasts on long-term demands for primary energy [84] published before. Demands for final energy obtained on the basis of such an analysis are in good agreement with the forecast based on the energy-GDP ratio. The results of the previous section (energy consumption of developing countries) are also in harmony with the forecast. A good agreement between the scales of produced primary energy obtained on the GEM-10R (see Chapters 7 and 8) and results for a medium (probable) scenario of IIASA–2 project [33] testify additionally to the satisfactory quality of this forecast.

3.4.2. Population

Many international and national organizations periodically make long-term forecasts of population growth for individual countries, regions and the world as a whole. The forecasts are usually based on a large set of assumptions relative to the future values of birth rate, death rate and migration. Due to their uncertainty the considered variants (more or less probable) are very often essentially different from each other.

The present work uses only one variant of the population growth forecast. It was chosen based on analysis of the latest forecasts of the following organisations: Department for Economic and Social Information and Policy Analysis of UN 1996 (UN96) [85] and 1998 (UN98) [86], US Census Bureau (USCB) [87], World Bank (WB)[21], International Institute for Applied System Analysis (IIASA) and World Energy Council 1995 (IIASA95) [33], IIASA 1996 (IIASA96) [88]. A summary of the characteristics of the above forecasts is given in Table 3.13 and Fig 3.6 presents their comparison by world population (in the cases when several forecast variants are given by one source the average or most probable variant is chosen).

In the 21st century, according to all the forecasts considered, world population growth rates will essentially decrease; by the end of the century its stabilization or even further drop is possible. The expected (probable) population number will make up 9–10 billion people by 2050 and 10–11.7 billion people by 2100. The latter (11.7 billion) is, apparently, overestimated as a result of ignoring new tendencies of decreasing birth rate in a number of countries that manifested themselves after the forecast had been made (1993–1994). In the remaining forecasts (in the average variant) 2100 values of world population differ by no more than 4%.

Table 3.13. Main characteristics of population forecasts

Characteristics	Forecast					
	UN96	UN98	USCB	WB	IIASA95	IIASA96
Publication year	1996	1998	1996	1998	1995	1996
Forecast horizon	2050	2150	2050	2150	2100	2100
Number of regions	All countries	9	All countries	All countries	11	13
Number of scenarios	3	7	1	1	1	27
World population, billion:						
2050	9.4	9.4	9.3	9.0	10.1	9.9
2100	–	10.4	–	10.0	11.7	10.4

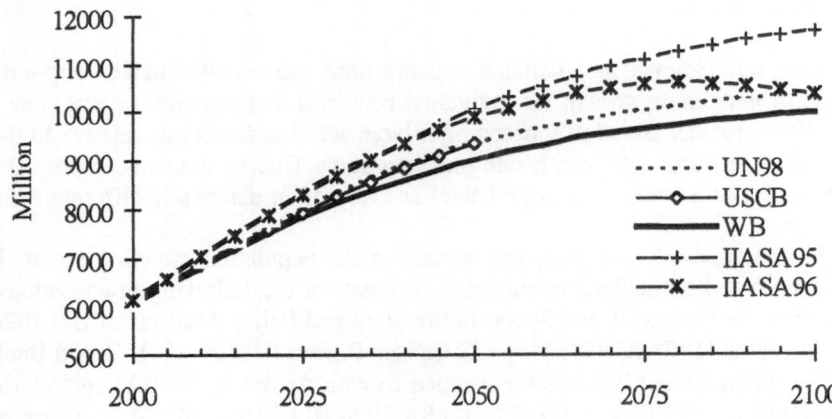

Figure 3.6. Comparison of the world population forecasts.

In further calculations we assumed the World Bank forecast [21] (slightly different from the average variants of the forecasts of the other organizations) for two reasons: firstly, it has an acceptable forecast horizon and secondly, it contains the data on individual countries, which allows their aggregation into the data on ten regions presented in the model GEM-10R (Table 3.14).

Table 3.14. Forecast of the population in 10 world regions, million people.

Year	Region										
	NA	EU	JK	AZ	SU	LA	ME	AF	CH	SA	World
1990	276	554	170	21	289	448	271	502	1284	1477	5292
2025	358	587	175	26	295	684	536	1092	1663	2409	7825
2050	370	565	162	26	294	800	685	1487	1752	2847	8988
2075	376	542	150	26	292	855	772	1776	1786	3081	9655
2100	379	541	148	26	296	874	803	1903	1814	3179	9962

An important peculiarity of the demographic situation in the 21st century is stabilisation of the population number in developed countries and in the former USSR region, so that practically all the growth (about 4 billion people) will fall on developing countries. Thus, their share in the world population will grow from 75% in 1990 to 86% in 2100 (Figure 3.7).

Taking into account that rise in life standard in these countries supposes increase in specific (per capita) energy consumption, it can be concluded that development scales and basic problems of world energy in the 21st century will depend, first of all, on developing countries.

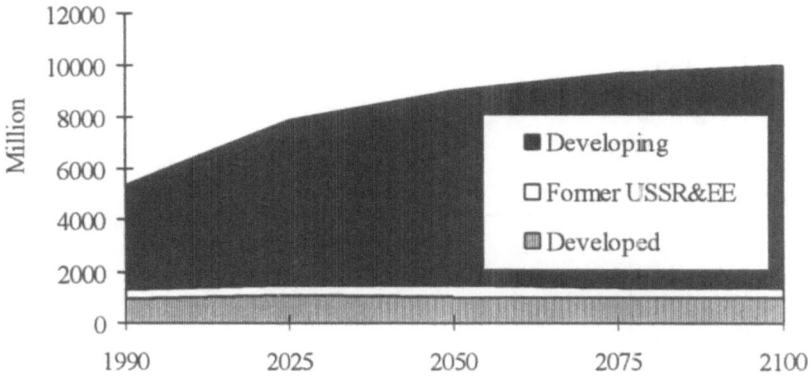

Figure 3.7. Forecast of population number in three world regions.

3.4.3. Economic growth

The hypothesis on possible approximate description of GDP growth rates using a single universal relation has been assumed as a basic principle of forecasting the economic development of all world regions in the 21st century. As historical experience shows (see Chapter 1) after reaching a definite level of economic development (per capita GDP is equal to US$1–4 thousand) a "splash" of annual GDP growth rates up to 7–8% is observed. Then the growth rates decline and at per capita GDP of US$10–20 thousand, when the country enters the category of industrially developed, amount to about 2% per year. In the future the economic growth rates of such a country are likely to slowly decline.

For approximation of the GDP growth rate dependence on the achieved economic development, the following relationship was used:

$$\beta = A\,x^k \exp(-Bx^l) + C\,x^m \exp(-Dx^n),$$

where β is a GDP growth rate, % per year, x is a specific GDP, US$ thousand/capita, and the coefficients are $A = 0.33$, $k = 6.5$, $B = 0.9$, $l = 1.45$, $C = 1.5$, $m = 0.25$, $D = 0.03$, $n = 0.95$. Figure 3.8 shows the extent of compliance of this approximation with the actual data [33] for 11 regions (according to the classification assumed in IIASA–2).

Based on this approximation, the forecast of the population number and initial conditions for 1990 (see Table 1.3) GDP was calculated for 2100 for 10 regions of the world (based on the classification of the GEM-10R model). For the former USSR region it was additionally assumed that its GDP would amount to 50% of the 1990 level by 2000 and would grow with the rate of 1–3% per year in 2005–2025.

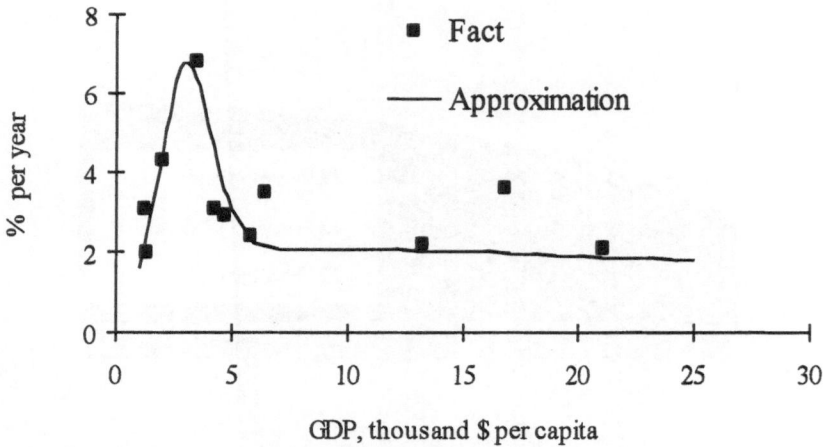

Figure 3.8. GDP growth rate versus GDP per capita.

The results of calculations (being in satisfactory agreement with the forecast of World Bank [21,33] and used for "*H*" variant) show that by the end of the 21st century the world gross product will increase 9 times compared to 1990 and make up US$236 trillion. An essential unevenness of economic development in the world regions will be retained to a great extent in the future as well (Table 3.15).

Table 3.15. Per capita GDP forecast (thousand 1990 US$ with regard to PPP) for the "*H*" variant (without assistance)

Year	Region										
	NA	EU	JK	AZ	SU	LA	ME	AF	CH	SA	World
1990	21.0	11.7	14.6	17.1	6.4	4.7	4.2	1.3	2.1	1.6	4.9
2025	30.0	21.8	27.4	26.6	4.7	7.1	5.9	1.2	6.6	4.8	8.0
2050	42.4	34.8	43.6	39.1	8.6	10.1	7.7	1.4	10.4	7.4	11.2
2075	56.3	50.9	62.4	53.9	14.3	15.6	11.4	2.5	16.9	11.3	16.1
2100	70.4	65.9	77.5	68.6	22.9	24.6	17.9	6.6	26.5	18.1	23.7

By 2100 there will be a 3–4-fold gap in the specific GDP value between the SU, LA and ME regions and the NA region, i.e. almost similar to what was in 1990. The CH and SA regions, which have reduced the gap from 10–13 to 3–4 times, will approach their level of development. At the same time by the end of the 21st century all the indicated regions will become industrially developed (in the modern understanding) and will top a 1990 level of the EU region. The AF region will still be the least economically developed. Per capita GDP of the region does not increase till the middle of the 21st century (low rates of economic growth are combined with the high rates of the population growth).

An important distinguishing feature of economic development in the 21st century is growing scales of the economy in developing countries: their total GDP that was 35% of world gross product in 1990 will exceed 50% in 2025 and reach 65% by 2100. In this connection possible assistance of developed countries to developing ones is likely to have a noticeable effect only in the nearest 2–3 decades.

3.4.4. High energy consumption ("H" variant)

The 1990 reports [37] on consumption of primary energy for production of final energy (electrical, thermal, mechanical and chemical), aggregated efficiencies of final energy production (see Chapter 1) and data on GDP were used to calculate energy-GDP ratios for individual forms of final energy in 10 world regions (see Table 3.16). Analysis of these data shows:
— developed countries (NA, EU, JK and AZ) have close values of energy intensity (slightly high energy intensity is characteristic of the NA region, particularly for mechanical energy);
— the former USSR region has the highest energy intensity for all forms of final energy;
— developing countries (LA, ME, AF, CH and SA) have higher thermal energy intensity compared to the developed regions but lower intensity of electrical, mechanical and chemical kinds of energy.

Table 3.16. Energy-GDP ratio by form of final energy in 1990 (GJ per thousand US$, with regard to PPP)

Energy form	Region										
	NA	EU	JK	AZ	SU	LA	ME	AF	CH	SA	World
Electrical	1.53	1.47	1.13	1.48	3.27	1.00	0.97	1.21	0.97	1.04	1.42
Thermal	3.87	3.29	2.64	2.92	11.27	5.12	4.74	8.27	5.54	3.65	4.51
Mechanical	0.88	0.48	0.32	0.39	0.98	0.29	0.30	0.27	0.23	0.22	0.51
Chemical	0.41	0.44	0.51	0.19	0.67	0.08	0.07	0.12	0.10	0.10	0.33

Relatively small differences in energy-GDP ratios of the regions compared to a great distinction in the level of economic development (per capita GDP) testifies to stability of this index and to its possible use for energy consumption forecasting.

Developed regions. It was supposed that in the period of 1990–2100 in the EU, JK and AZ regions energy intensity by individual energy forms will decrease with the rate (%/year) of: 1.3 for electrical energy, 1.5 for thermal energy, 1 for mechanical and 0.8 for chemical. In the NA region the rate of energy intensity decrease for electrical and thermal energy is assumed equal to 1 and 1.3% per year respectively. Such rates of energy intensity decrease are in compliance with the forecasts of IIASA-2 [33] (between the values of indices for scenarios B and C).

Developing regions and the former USSR. In the 21st century economic development of these regions will reach the (1990) level of per capita GDP of the EU region (US$ 11.7 thousand/capita): CH — in 2055, LA — in 2060, SU — in 2065, ME and SA— in 2075.

Energy-GDP ratios of the indicated regions for individual forms of final energy were supposed to change (decrease or increase) with the permanent rate and by a definite time to reach the energy intensity level of the EU region, not 1990 but 2005 level, i.e. somewhat lesser in magnitude: 1.2 — for electrical energy, 2.6 — for thermal, 0.41 — for mechanical and 0.39 GJ/thousand US$ for chemical energy. To take into account the specific features of the former USSR region (the need for considerable amounts of thermal energy for heating) energy intensity in 2065 is assumed twice as much, i.e. 5.2 GJ/thousand US$. Besides it was supposed that in parallel with a two-fold decrease of GDP by 2000, energy intensity for all energy forms increases almost two-fold.

In the succeeding period the LA, ME, CH and SA regions will develop like the EU region and the former USSR region — the NA region (the energy-GDP ratio decreases at the same rate).

The results of the energy consumption forecast are presented in Tables 3.17, 3.18 and in Figs 3.9, 3.10. Some growth of per capita electrical energy consumption is expected in developed countries in the period of up to the mid-21st century (up to 35–36 GJ/capita/yr in NA and 23–26 GJ/capita/yr in EU, JK and AZ) with subsequent insignificant decreases. Per capita electric energy consumption in the SU region in the first half of the 21st century will keep on decreasing (up to 15–16 GJ/capita/yr) with subsequent restoration of 1990 level (20 GJ/capita/yr) by the end of the 21st century. Gradual growth of electrical energy consumption up to 15–17 GJ/capita/yr in 2100 is expected in the LA, ME, CH, SA regions, which corresponds to the current level of electrical energy consumption in Europe and Japan. By 2100 the AF region will reach only half of this level.

Per capita thermal energy consumption in the NA and AZ regions will decrease to 65 and 38 GJ/capita/yr by 2100 respectively. In the EU and JK regions specific thermal energy consumption will grow till the mid-21st century and reach 46 GJ/capita/yr, then decrease to 39–41 GJ/capita/yr by 2100.

In the former USSR region, a decrease of per capita thermal energy consumption in the first half of the 21st century (up to 63 GJ/capita/yr) gives way to growth, and energy consumption will reach 76 GJ/capita/yr by the end of the century. In developing countries (except for AF) a permanent growth of per capita thermal energy consumption is expected up to the magnitude of 32–34 GJ/capita/yr in 2100. In the AF region in the first quarter of the 21st century, per capita thermal energy consumption will decrease due to a large population growth. An evident energy consumption growth starts only in the second half of the century and reaches 17 GJ/capita/yr) by 2100.

Per capita consumption of mechanical and chemical energy will increase in all the regions except for SU where it will decrease in the first quarter of the 21st century; subsequent growth will lead to restoration of the 1990 level by 2050 for chemical and by 2100 for mechanical energy.

Table 3.17. Final energy consumption, GJ/capita/yr (*"H"* variant)

Year	NA	EU	JK	AZ	SU	LA	ME	AF	CH	SA	World
					Electrical						
1990	32.0	17.1	16.5	25.3	20.9	4.7	4.1	1.6	2.0	1.7	7.0
2025	32.1	20.2	19.6	24.8	16.1	7.8	6.3	1.4	7.1	5.3	8.6
2050	35.3	23.2	22.5	26.3	15.6	11.8	8.9	1.7	12.1	8.5	11.0
2075	36.5	24.5	23.2	26.1	16.1	15.3	14.2	3.0	15.3	13.4	13.9
2100	35.6	22.9	20.8	24.0	20.0	17.4	16.1	8.0	17.3	15.4	16.0
					Thermal						
1990	81.2	38.4	38.5	50.0	72.1	24.1	19.9	10.8	11.6	5.9	22.2
2025	73.3	42.3	42.5	45.7	59.7	25.7	21.7	6.7	24.1	15.2	24.4
2050	74.7	46.2	46.4	46.1	62.7	28.3	24.0	6.1	28.0	21.3	26.6
2075	71.5	46.3	45.5	43.6	65.6	31.5	29.6	8.5	31.6	29.4	30.1
2100	64.6	41.1	38.8	38.0	75.5	34.0	32.1	17.2	33.9	32.1	32.9
					Mechanical						
1990	18.4	5.6	4.7	6.7	6.3	1.4	1.3	0.4	0.5	0.4	2.5
2025	18.5	7.4	6.3	7.3	4.9	2.5	2.0	0.4	2.1	1.4	3.0
2050	20.3	9.2	7.8	8.4	4.9	4.0	2.9	0.5	4.1	2.6	4.0
2075	21.0	10.4	8.6	9.0	5.1	5.6	4.8	1.0	5.7	4.7	5.3
2100	20.4	10.5	8.3	8.9	6.4	6.9	5.9	2.8	6.9	5.9	6.4
					Chemical						
1990	8.6	5.1	7.4	3.3	4.3	0.4	0.3	0.2	0.2	0.2	1.6
2025	9.2	7.2	10.5	3.9	4.0	1.2	0.8	0.2	1.4	0.8	2.1
2050	10.7	9.4	13.7	4.7	4.8	2.8	1.8	0.3	3.6	1.8	3.2
2075	11.6	11.3	16.0	5.3	5.7	4.8	4.3	0.7	5.5	4.0	4.7
2100	11.9	11.9	16.3	5.5	7.5	6.2	5.6	2.6	7.1	5.2	6.0
					Total						
1990	140.2	66.2	67.2	85.4	103.6	30.5	25.5	12.8	14.3	8.1	33.2
2025	133.1	77.2	78.9	81.7	84.7	37.2	30.9	8.7	34.7	22.6	38.0
2050	141.1	88.0	90.3	85.5	88.0	47.0	37.7	8.6	47.8	34.1	44.7
2075	140.7	92.6	93.4	84.0	92.6	57.3	52.9	13.2	58.0	51.6	54.1
2100	132.5	86.4	84.2	76.4	109.3	64.3	59.6	30.6	65.2	58.6	61.3

World total consumption of final energy (see Table 3.18, Figure 3.10) will increase almost 3.5 times by the end of the 21st century and will be characterised by a considerable increase in the fraction of the developing regions (from 33% in 1990 to 76% in 2100).

Table 3.18 Final energy consumption, million TJ/yr ("*H*" variant)

Year						Region					
	NA	EU	JK	AZ	SU	LA	ME	AF	CH	SA	World
						Electrical					
1990	8.8	9.5	2.8	0.5	6.1	2.1	1.1	0.8	2.6	2.5	36.8
2025	11.5	11.9	3.4	0.6	4.7	5.3	3.4	1.5	11.9	12.7	67.1
2050	13.1	13.1	3.6	0.7	4.6	9.4	6.1	2.5	21.2	24.1	98.5
2075	13.7	13.3	3.5	0.7	4.7	13.1	10.9	5.4	27.4	41.4	134.1
2100	13.5	12.4	3.1	0.6	5.9	15.2	13.0	15.3	31.4	49.0	159.3
						Thermal					
1990	22.4	21.3	6.5	1.1	20.8	10.8	5.4	5.4	14.9	8.7	117.2
2025	26.3	24.8	7.4	1.2	17.6	17.6	11.6	7.3	40.1	36.6	190.6
2050	27.6	26.1	7.5	1.2	18.4	22.7	16.5	9.1	49.1	60.5	238.7
2075	26.9	25.1	6.8	1.1	19.1	27.0	22.9	15.1	56.3	90.7	291.0
2100	24.4	22.2	5.7	1.0	22.3	29.7	25.7	32.7	61.5	102.1	327.5
						Mechanical					
1990	5.1	3.1	0.8	0.1	1.8	0.6	0.3	0.2	0.6	0.5	13.2
2025	6.6	4.3	1.1	0.2	1.5	1.7	1.1	0.4	3.4	3.3	23.7
2050	7.5	5.2	1.3	0.2	1.4	3.2	2.0	0.7	7.2	7.3	36.0
2075	7.9	5.7	1.3	0.2	1.5	4.8	3.7	1.7	10.1	14.6	51.4
2100	7.7	5.7	1.2	0.2	1.9	6.0	4.7	5.3	12.5	18.6	63.9
						Chemical					
1990	2.4	2.8	1.3	0.1	1.2	0.2	0.1	0.1	0.3	0.2	8.6
2025	3.3	4.2	1.8	0.1	1.2	0.8	0.4	0.2	2.3	1.9	16.4
2050	4.0	5.3	2.2	0.1	1.4	2.3	1.2	0.5	6.3	5.1	28.4
2075	4.4	6.1	2.4	0.1	1.7	4.1	3.3	1.3	9.8	12.4	45.6
2100	4.5	6.4	2.4	0.1	2.2	5.4	4.5	4.9	12.8	16.6	59.9
						Total					
1990	38.7	36.7	11.4	1.8	29.9	13.7	6.9	6.4	18.4	11.9	175.9
2025	47.7	45.3	13.8	2.1	25.0	25.5	16.5	9.5	57.6	54.6	297.6
2050	52.1	49.8	14.6	2.2	25.9	37.6	25.8	12.7	83.8	97.0	401.5
2075	52.9	50.2	14.0	2.2	27.0	49.0	40.8	23.5	103.6	159.0	522.1
2100	50.1	46.7	12.5	2.0	32.3	56.2	47.9	58.2	118.2	186.4	610.5

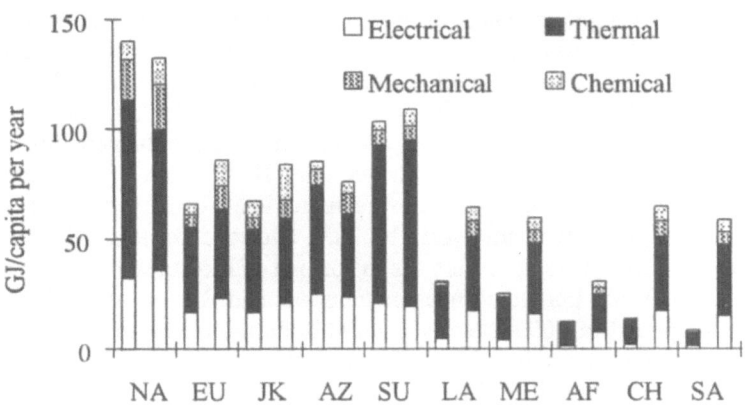

Figure 3.9. Final energy consumption, GJ/capita/yr (left column — 1990, the right one — 2100).

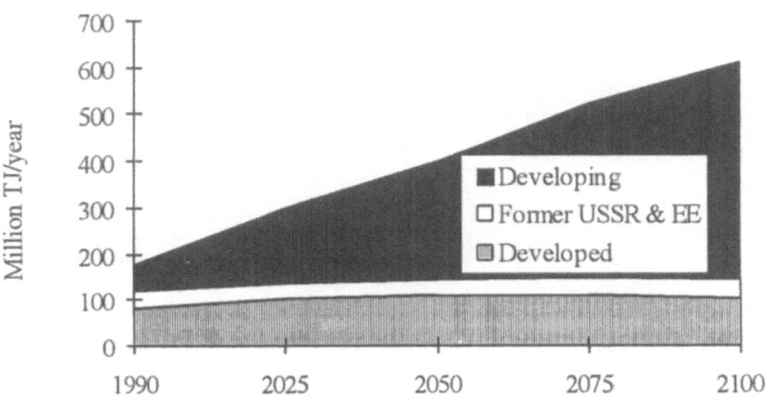

Figure 3.10. Total consumption of final energy in the three world regions.

3.4.5. Low energy consumption ("*L*" variant)

This variant supposes partnership and cooperation between states in the 21st century to gradually pass to sustainable development (decrease in economic inequality and reduction of negative anthropogenic impact on the environment). Two factors were taken into account in the forecast of economic development and energy consumption: 1) financial assistance of developed to developing countries and 2) more intensive decrease in energy-GDP ratio of developing countries.

Potential size of the assistance from developed to developing countries was determined based on the indices discussed at the conference in Rio de Janeiro: about 0.7 % of GDP of developed countries or US$ 100–125 billion per year [50]. The calculations assumed that each of the developed regions (NA, EU, JK and AZ) will annually allocate about 35% of their yearly GDP growth for assistance to developing countries during the period of 2000–2025 and 20% during the period of 2025–2050, which will total US$ 95–160 billion per year. Distribution of this assistance was modeléd based on the objective to ensure approximate equality of per capita GDP of the remaining regions of the world by 2050 and to make this index closer to the 1990 level of the EU region. Here the major part of the assistance is rendered to the most economically underdeveloped regions AF and SA: 60 and 30% respectively during the period of 2000–2025, 45 and 40% during the period of 2025–2050. No assistance is rendered to the CH and LA regions.

Thus, per capita GDP of the SU, ME, AF and SA regions increases compared to the previous variant (particularly sharply in the AF region); unevenness of its distribution is leveled out, though there remains approximately a three-fold lagging of the developing regions behind the developed ones (Table 3.19).

Table 3.19. Per capita GDP (thousand 1990 US$ with regard to PPP), the variant with the assistance rendered by developed to developing countries

Year	Region										
	NA	EU	JK	AZ	SU	LA	ME	AF	CH	SA	World
1990	21.0	11.7	14.6	17.1	6.4	4.7	4.2	1.3	2.1	1.6	4.9
2025	26.0	18.6	23.6	22.9	5.5	7.1	6.2	5.4	6.6	5.3	8.2
2050	35.1	27.9	35.8	32.0	9.4	10.1	8.7	7.8	10.4	8.2	11.7
2075	48.7	42.7	53.6	46.1	15.6	15.6	12.9	10.8	16.9	12.5	17.2
2100	63.1	57.7	69.1	60.9	24.7	24.6	20.2	16.6	26.5	19.8	25.5

It is interesting to note (comparing Tables 3.15 and 3.19) that, when rendering assistance to developing countries, per capita GDP for the world as a whole grows by 7–8% in the second half of the century, while in developed countries it decreases by 10–15%. Hence, such an assistance increases world GDP and is a progressive and effective measure in terms of the whole of humanity. However taking into account still existing large differences in economic development (about a 3-fold difference in the magnitude of per capita GDP and lagging by 70–80 years in time) it seems reasonable to consider the possibilities of increasing the assistance size.

Energy consumption (Tables 3.20 and 3.21) was calculated using the same approach as in "*H*" variant.

Energy-GDP ratio for all final energy forms in the regions LA, ME, AF, CH and SA in 2050 was set equal to energy intensity of the EU region in 2025, i.e. 0.93 for electrical energy, 1.9 — for thermal, 0.34 — for mechanical, 0.33 GJ/thousand dollars — for chemical energy; for the SU region the previous values of indices were retained, since expert analysis showed low probability of an essential decrease in per capita energy consumption compared to "*H*" variant.

Table 3.20. Final energy consumption, GJ/capita/yr ("*L*" variant)

Year					Region						
	NA	EU	JK	AZ	SU	LA	ME	AF	CH	SA	World
					Electrical						
1990	32.0	17.1	16.5	25.3	20.9	4.7	4.1	1.6	2.0	1.7	7.0
2025	27.9	17.2	16.9	21.3	15.2	6.9	5.8	5.7	6.2	5.1	8.3
2050	29.3	18.7	18.5	21.5	11.21	9.5	8.0	7.4	9.5	7.6	10.1
2075	31.6	20.6	20.0	22.4	14.5	10.6	8.4	7.4	11.1	8.3	10.9
2100	31.9	20.0	18.5	21.3	17.9	12.0	9.5	8.2	12.6	9.5	11.9
					Thermal						
1990	81.2	38.4	38.5	50.0	72.1	24.1	19.9	10.8	11.6	5.9	22.2
2025	63.7	36.0	36.6	39.4	59.5	20.8	17.2	19.0	20.2	13.3	22.8
2050	61.9	37.1	38.1	37.7	50.1	19.7	16.7	15.0	20.7	15.8	21.9
2075	61.9	38.9	39.1	37.2	60.1	20.8	16.8	14.2	22.9	16.6	22.5
2100	57.8	36.0	34.5	33.7	68.7	22.4	18.1	15.0	24.6	18.0	23.3
					Mechanical						
1990	18.4	5.6	4.7	6.7	6.3	1.4	1.3	0.4	0.5	0.4	2.5
2025	16.1	6.3	5.4	6.3	4.9	2.3	2.1	1.7	1.9	1.5	3.0
2050	16.8	7.3	6.4	6.9	3.9	3.4	3.1	2.7	3.6	2.8	4.0
2075	18.2	8.8	7.4	7.7	5.1	4.0	3.6	2.9	4.6	3.4	4.6
2100	18.3	9.2	7.4	7.9	6.3	4.9	4.4	3.5	5.6	4.1	5.3
					Chemical						
1990	8.6	5.1	7.4	3.3	4.3	0.4	0.3	0.2	0.2	0.2	1.6
2025	8.0	6.1	9.0	3.4	4.0	1.3	1.0	1.1	1.4	1.1	2.1
2050	8.9	7.5	11.2	3.8	3.9	3.2	2.7	2.5	3.6	2.7	3.6
2075	10.1	9.5	13.7	4.5	5.2	4.0	3.3	2.8	4.8	3.4	4.4
2100	10.7	10.4	14.5	4.9	6.8	5.2	4.2	3.5	6.1	4.3	5.4
					Total						
1990	140.2	66.2	67.2	85.4	103.6	30.5	25.5	12.8	14.3	8.1	33.2
2025	115.7	65.7	67.9	70.4	83.6	31.2	26.0	27.5	29.7	21.0	36.3
2050	117.0	70.6	74.1	70.0	69.1	35.8	30.5	27.5	37.4	28.9	39.6
2075	121.7	77.7	80.2	71.8	85.0	39.5	32.1	27.3	43.4	31.7	42.3
2100	118.7	75.6	75.0	67.8	99.6	44.6	36.2	30.2	48.9	36.1	45.9

Thus, energy consumption in most world regions decreases (in the AF region increases by the middle of the 21st century compared to the previous variant), the total energy consumption in 2100 appears to be 25% lower and the fraction of developing countries makes up 72% instead of 76% in "*H*" variant (Figures 3.11, 3.12).

Thus, "*L*" variant is characterised by faster equalisation of economic development levels of the regions, decrease in total energy consumption and correspondingly negative impact of energy on the environment. Since the scales of economic and technological assistance that were envisaged in this variant do not exceed really possible scales, the variant can be considered as a first approximation description of gradual change in the existing tendencies and of transition to sustainable development.

Table 3.21. Final energy consumption, million TJ/yr (*"L"* variant)

Year	NA	EU	JK	AZ	SU	LA	ME	AF	CH	SA	World
						Region					
					Electrical						
1990	8.8	9.5	2.8	0.5	6.1	2.1	1.1	0.8	2.6	2.5	36.8
2025	10.0	10.1	2.9	0.6	4.5	4.7	3.1	6.2	10.3	12.4	64.8
2050	10.8	10.5	3.0	0.6	3.3	7.6	5.4	11.0	16.7	21.5	90.6
2075	11.9	11.2	3.0	0.6	4.2	9.1	6.5	13.1	19.9	25.7	105.2
2100	12.1	10.8	2.7	0.6	5.3	10.5	7.7	15.6	22.8	30.3	118.4
					Thermal						
1990	22.4	21.3	6.5	1.1	20.8	10.8	5.4	5.4	14.9	8.7	117.2
2025	22.8	21.2	6.4	1.0	17.6	14.2	9.2	20.8	33.5	32.1	178.8
2050	22.9	21.0	6.1	1.0	14.7	15.7	11.5	22.3	36.2	45.1	196.5
2075	23.3	21.1	5.9	1.0	17.5	17.8	13.0	25.3	40.9	51.2	216.9
2100	21.9	19.5	5.1	0.9	20.3	19.6	14.5	28.6	44.7	57.3	232.4
					Mechanical						
1990	5.1	3.1	0.8	0.1	1.8	0.6	0.3	0.2	0.6	0.5	13.2
2025	5.8	3.7	0.9	0.2	1.4	1.5	1.1	1.8	3.2	3.6	23.3
2050	6.2	4.2	1.0	0.2	1.2	2.7	2.1	4.0	6.4	8.0	36.0
2075	6.8	4.7	1.1	0.2	1.5	3.4	2.8	5.1	8.1	10.4	44.2
2100	6.9	5.0	1.1	0.2	1.8	4.3	3.5	6.6	10.1	13.2	52.7
					Chemical						
1990	2.4	2.8	1.3	0.1	1.2	0.2	0.1	0.1	0.3	0.2	8.6
2025	2.9	3.6	1.6	0.1	1.2	0.9	0.6	1.3	2.3	2.5	16.8
2050	3.3	4.3	1.8	0.1	1.1	2.6	1.9	3.7	6.3	7.6	32.6
2075	3.8	5.1	2.1	0.1	1.5	3.5	2.5	5.0	8.5	10.3	42.4
2100	4.0	5.6	2.1	0.1	2.0	4.5	3.4	6.7	11.1	13.8	53.5
					Total						
1990	38.7	36.7	11.4	1.8	29.9	13.7	6.9	6.4	18.4	11.9	175.9
2025	41.5	38.6	11.9	1.8	24.7	21.4	14.0	30.1	49.3	50.6	283.7
2050	43.2	39.9	12.0	1.8	20.3	28.6	20.9	40.9	65.6	82.3	355.6
2075	45.7	42.2	12.0	1.9	24.8	33.8	24.8	48.5	77.4	97.6	408.7
2100	44.9	40.9	11.1	1.8	29.5	39.0	29.1	57.5	88.6	114.6	456.9

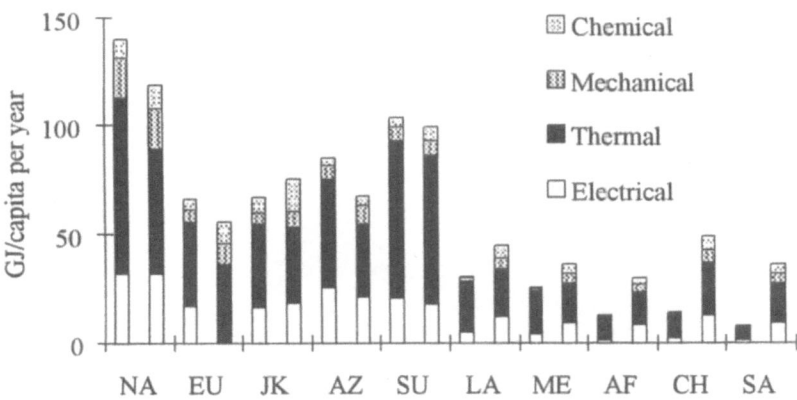

Figure 3.11. Final energy consumption, GJ/capita per year (left column — 1990, right column— 2100); "*L*" variant.

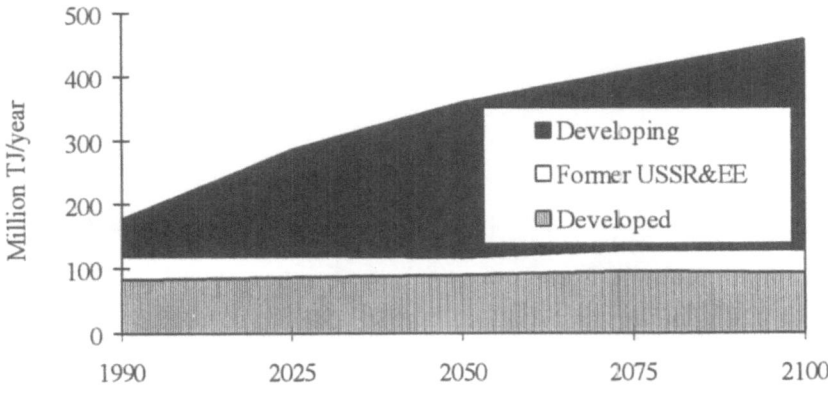

Figure 3.12. Total consumption of final energy in the three world regions, "*L*" variant.

Chapter 4

ENERGY RESOURCES

Evaluation of constraints on energy resources have invariably ranked high in world energy studies. Many known specialists continue to participate in this work (M.Grenon, S.Masters, B.Dessous, F.Farabod et al. abroad and M.S.Modelevsky, N.V.Chersky, A.S.Astakhov, Yu.V.Sinyak, I.A.Bashmakov et al. in Russia). Resource problems are regularly discussed at congresses of the World Energy Council and oil, gas and coal congresses. In the last two decades international symposia on renewable energy sources have been held frequently dealing with the problems of assessing the appropriate resources. These resource studies occupy a highly important place in the activity of many authoritative organisations, in particular IIASA and IAEA.

Estimates of energy resources for different countries and regions of the world are given in numerous publications. They are constantly supplemented and corrected. The analysis shows that the presented data are often inconsistent with each other, the divergences being pronounced. This is explained, in particular, by differences in the degree to which technical, environmental, economic and other constraints are taken into account. It is especially typical of estimates of renewable energy resources.

The key objectives of this chapter are:

a) generalisation of available published data on the most critical energy resources and their cost analysis;

b) choice of a reasonable degree of aggregation for correct representation of each resource in the GEM-10R model;

c) estimation of constraints on energy resources for the 10 world regions by the accepted classification.

Data from numerous publications have been used in the work; the most essential information is cited in the list of publications. A lot of valuable information on energy resources for individual countries was extracted from the proceedings of such congresses as WEC and from special issues of such journals as *"World Oil"*, *"Oil and Gas"*, *"BP Statistical Review of World Energy"*, *"UN Energy Statistics Yearbook"*, *"WEC Survey of Energy Resources"*, *"Energy Policy"*, *"Water Power and Dam Construction"*, *"Wind Engineering"*, *"Solar Energy"*, *"Geothermics"*, *"Annual Review of Energy"*, etc. Results of resource studies performed directly by the authors have also been used [89, 90].

4.1. Methodological aspects

Energy resources here are taken to mean natural resources of the Earth and space that can be utilised in a foreseeable future to produce electrical, thermal, mechanical and chemical energy. These are primary energy resources. Their use routinely results in waste, i.e. secondary energy resources (low-grade thermal energy, for example), which may be utilised later, whenever economically beneficial. In some cases such resources require preliminary processing (for example, plutonium extraction from spent nuclear fuel). Organic waste products of agriculture, industry (primarily, forest and woodworking industries) and household life are sizeable sources of secondary energy.

Primary energy resources may be classified by the following criteria:

1) location — resources of the lithosphere (fossil organic fuels, biomass, nuclear fuel, geothermal energy), the hydrosphere (water power of rivers, energy of tides and waves, biomass of hydrobiota, water-dissolved uranium and tritium), the atmosphere (wind energy, solar energy at the earth surface), space (solar energy beyond the Earth, resources of helium–3 on the Moon and other planets of the Solar System);

2) reproducibility by nature — reproducible (renewable) and virtually non-reproducible (non-renewable) resources in a foreseeable time horizon;

3) exhaustibility — limited (exhaustible) and virtually unlimited (inexhaustible) resources;

4) form of original (natural) energy — organic fuels (combustible materials, chemical energy), nuclear (fissile materials) and thermonuclear fuels (fusion materials), renewable inorganic energy forms (mechanical, electromagnetic, thermal, etc.).

Reserves of fossil organic fuels (oil, gas and coal) are limited and virtually non-reproducible resources. Energy resources of sun, wind, oceans and rivers, biomass and thermal sources belong to a category of renewable and reproducible resources. Renewable energy sources, except for biomass, are inorganic resources. A salient feature of using reserves of fossil organic and nuclear fuels is the necessity of their preliminary multistage preparation to exploitation: discovery as a result of prospecting, determination of volumes and conditions of occurrence in the course of preliminary and detailed exploration, stripping of resources, etc.

Traditionally, reserves of fossil organic and nuclear fuels are classified on the basis of the following characteristics: accessibility, reliability, geological conditions of occurrence, natural quality, technical recoverability, economic efficiency of use, readiness for exploitation, etc. [91, 92]. McKelvey's classification of mineral resources is the most widely accepted in the world [48, 91–93]. McKelvey's scheme is based on two classification characteristics: reliability and economic efficiency of reserves (Figure 4.1). In terms of the reliability (exploration) level the following groups of reserves are differentiated:

— identified reserves consisting of measured, indicated and inferred reserves;

— undiscovered reserves including hypothetical and assumed or speculative reserves.

As to economic efficiency, the reserves are divided into economic and subeconomic which in turn are subdivided into paramarginal and submarginal. Subeconomic reserves are not profitable now, but can become profitable in the future. The cost of resource extraction rises with transition from discovered to undiscovered reserves, from measured to indicated and further to hypothetical and speculative ones.

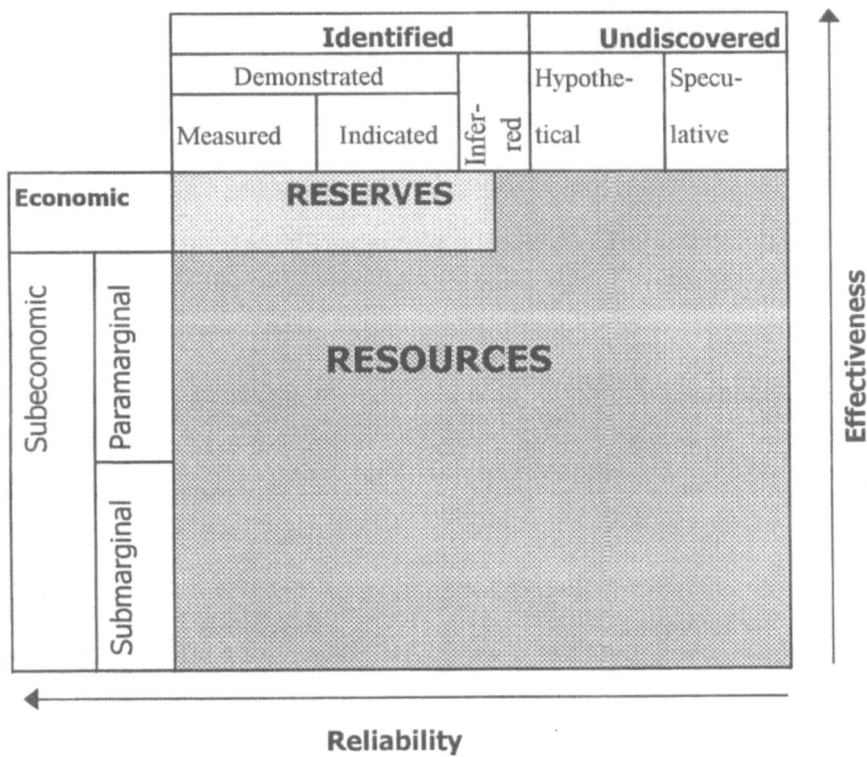

Figure 4.1. McKelvey's classification of mineral resources.

Comparison between the Western and Russian (Soviet) classifications of energy resources is shown in Table 4.1 [94]. According to the Soviet classification, reserves are denoted by letters from A to D with indices from 1 to 3. In degree of decreasing reliability, the reserves are divided into categories A, B, C_1 and C_2 (including forecasted ones). The balance reserves include commercial reserves of categories $A+B+C_1$ and preliminarily estimated reserves of category C_2. Resources of categories $C_3+D_1+D_2$ are unexplored.

Renewable energy forms (sources) are divided sometimes into conventional (hydroenergy and biomass) and unconventional. Basic unconventional renewable energy sources comprise solar radiation, kinetic energy of air mass transport (wind), internal thermal (geothermal) energy of the Earth, gravitation forces of the Sun, Moon and Earth causing tidal phenomena in the ocean. On the other hand, renewable energy sources can be classified into organic (biomass of land and water reservoirs) and inorganic. Renewable inorganic energy forms (sources) are characterised by a wide variety (hydroenergy, solar energy, wind energy, geothermal energy, etc.).

Theoretically, resources of the majority of primary energy forms on the Earth are huge. However, only some part of them can be extracted by current and prospective technologies for further utilisation. Actually the economically efficient (or commercial) resources are even fewer because of additional constraints: economic, ecological, social, legislative, etc.

Table 4.1. Correspondence between the Western and Russian resource classifications

Western classification	Russian classification
Identified	Balance $(A + B + C_1 + C_2)$
Measured	Explored $(A + B)$
Indicated	C_1 and C_2 in known fields
Inferred	C_2 in new areas and C_3 in known fields
Undiscovered	Resources $(C_3 + D_1 + D_2)$
Hypothetical	Perspective $(C_3 + D_1)$
Speculative	Forecasted (D_2)

In accordance with the above the energy resources are usually divided into theoretical, technically possible (extractable) and economic. Therefore, it is customary to distinguish theoretical (gross or natural), technical and economic potentials of energy resources in publications.

Theoretical (gross or natural) potential is the maximum possible scale of using an energy resource of the specified form (in other words, it is the total energy inherent in the specified energy resource).

Technical potential is a part of the theoretical potential that can be practically utilised at the existing level of science and technology development. The technical potential is substantially (as a rule, 2–3 orders of magnitude and more) lower than the theoretical (gross) one, but it steadily increases with equipment improvement and introduction of new technologies.

Economic potential is a part of the technical potential, whose development is economically justified at the existing level of prices of equipment and materials, manpower and competing energy resources. It also changes in time with variation of technical and economic characteristics of extraction technologies and the state of world prices of alternative energy resources.

Theoretical resources of fossil organic fuels progressively increase owing to first, improvement of exploration methods and second, expansion of territories with intensive prospecting by advanced methods.

In general, the following are the main factors determining the volume of the economic part of energy resources:

— discovery of new fields with favourable geological conditions (for non-renewable energy forms);

— development of technologies for energy resources extraction, in particular improvement of their economic characteristics;

— economic, social and other constraints on extraction of a specified resource in the region.

Hence, the economic part of energy resources is not constant. Under the influence of the indicated factors it can vary in either direction. Therefore, it is expedient to repeat periodically resource investigations. Besides, the cost characteristic of the resource $C = f(R)$, i.e. dependence of its cost on an annual extraction for renewable resources and on the cumulative one for non-renewable resources, can also vary.

These principles are illustrated in Figure 4.2. Here the resource volumes are denoted as follows: R_{TH} — theoretical, R_T — technical (extractable), R_E — total economic, R_C — competitive and R_0 — used. Change of the cost characteristic has the greatest effect on the competitive resource part. It increases (point R''_C) with the use of new, more effective extraction technologies and decreases (point R'_C) with appearance of additional expenditures (for example, social expenditures). In Figure 4.2 the letter C_a denotes a system (regional) cost of the alternative (marginal) energy form. The value of R_C is reduced with decrease of C_a (line AB descends) and R_C increases with its growth (line AB ascends).

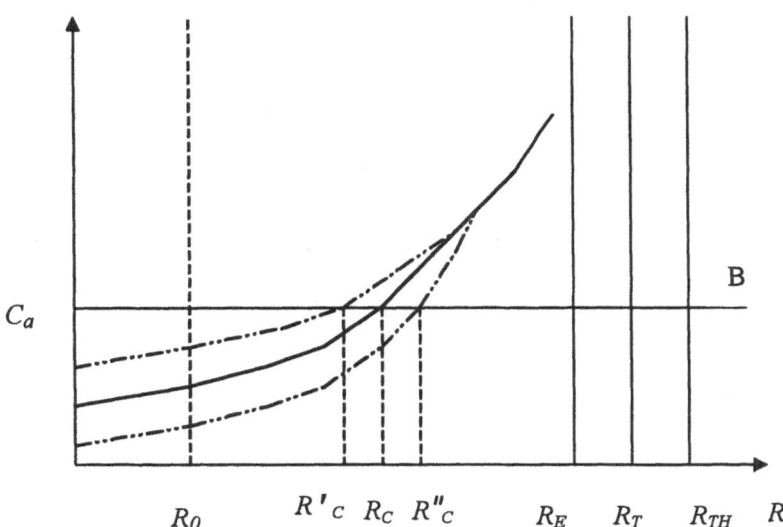

Figure 4.2. Cost analysis of energy resources.

To be applied in the GEM-10R model the energy resources (their economic part) are represented as follows (Figure 4.3):

$$R_E = R_0 + \sum_{i=1}^{n} \Delta R_i$$

The cost characteristic of each component of the resource has the form

$$C_i = C_{i-1} + \Delta C_i (R),$$

where the cost increase ΔC_i of a resource takes discrete values as a function of its extraction volumes.

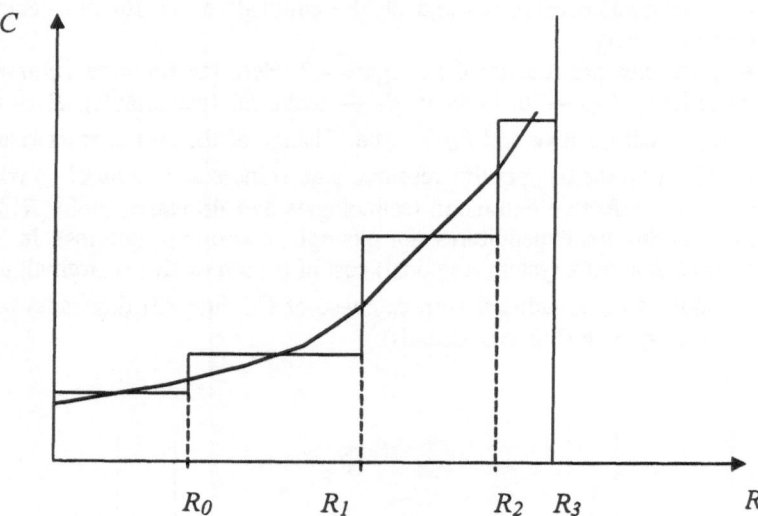

Figure 4.3. Representation of energy resources in GEM-10R model.

The number of resource gradations (n) by cost should not be large in order not to increase the model dimension. However, with small n the model can turn out to be insufficiently precise (because of unnecessary discreteness in the cost of competing resources). Experience shows that the value of n is determined by the cost range of the resource (from 5 to 8). It is not always obligatory to distinguish the component R_0, but in some cases it should be done (for example, for hydroenergy).

The studies performed have shown that the form of function $C=f(R)$ in Figure 4.2 is typical of essentially all primary energy forms.

The authors applied in their studies primarily the "bottom-up" method. It supposes at first an estimation of energy resources for each country and then their summation for the regions. This approach is rather laborious, but it allows more strictly consideration of additional constraints on resources.

4.2. Fossil fuel resources

The GEM-10R model applies a cost classification of fossil fuel resources that is based on the data from [48, 93]. It is presented in Table 4.2 (the average values of extraction cost ranges given in [48, 93] are taken there).

Table 4.2. Fossil fuel extraction cost, US$/GJ

Resource	Category							
	1	2	3	4	5	6	7	8
Oil	1.8	2.8	3.9	5.4	6.5	8.0	10.1	19.6
Gas	2.0	2.9	4.3	5.5	6.4	7.6	9.0	18.1
Coal	1.1	1.6	2.0	2.3	2.8	3.4	4.1	5.5

Oil. Tables 4.3 and 4.4 present estimates of extractable resources of conventional and unconventional oil with separation of discovered, probable (with a 50% probability of discovery) and additionally possible resources. They were obtained from the data of S. Masters and WEC for individual countries [95, 96] and then corrected in terms of other sources. With a measure of optimism the estimates may be assumed equal to the economic part of these resources. For oil (as well as for natural gas) this is admissible, since usually there are no strict (insuperable) additional constraints on production of "extractable resources".

Table 4.3. Conventional oil resources, million TJ

Region	Resources			
	Discovered	Probable (50% of discovery probability)	Additionally possible for discovery	Total
NA	320	468	959	1747
EU	165	160	131	456
JK	0	0	0	0
AZ	14	28	17	59
SU	457	577	594	1628
LA	707	451	423	1581
ME	4007	805	1091	5903
AF	126	160	86	372
CH	137	274	74	485
SA	106	160	126	392
World	6039	3083	3501	12623

The structure of these oil resources in terms of production cost may be taken as in [97]. Here and further (unless otherwise specified) the costs are expressed in 1990 US dollars. The estimates of oil resources by cost category for all the 10 regions that were obtained by us in 1994 from this information are presented in [89].

Later on H.-H. Rogner made an assessment of the world oil resources [48, 93]. The estimates are very optimistic with respect to oil reserves of expensive cost categories that comprise mainly resources of unconventional oil (shale, oil sand, etc.) and the remainder of oil reserves in old fields (that can not be extracted by traditional technologies). These estimates with division of resources into eight cost categories were applied in the described studies.

Table 4.4. Resources of unconventional oil (heavy oil, shale, tar sand and natural bitumen), million TJ

Region	Resources		
	Discovered	Probable and additionally possible for discovery	T o t a l
NA	22	9962	9984
EU	12	31	43
JK	0	0	0
AZ	153	1360	1513
SU	84	2268	2352
LA	15	979	994
ME	265	1151	1416
AF	2	12	14
CH	11	20	31
SA	6	18	24
World	570	15801	16371

Table 4.5 illustrates quantitative values of oil resources (both conventional and unconventional) that were used in the runs on GEM-10R, with division into cost categories, corresponding to Table 4.2. According to these data the share of very expensive oil (categories 6–8) accounts for some 82% of its world reserves. Table 4.5, like the next analogous Tables for natural gas, coal and uranium, does not include resources extracted by the year 1990.

The following are the richest oil regions with an acceptable extraction cost (categories 1–5) (TJ/capita): ME — 31.8, SU — 11.5, AZ — 11.4, NA — 7.2 and LA — 5.9 at an average world value equal to 3.9 (Table 4.6). The JK, SA and CH regions are least provided with oil (<1.0 TJ/capita).

Table 4.7 presents information on the current level of oil production and consumption in these regions [98]. The world resources of the first five categories (see Table 4.5) seem to be sufficient approximately for 130 years at production volumes of the year 2000.

Natural gas. Rather reliable estimates of extractable resources of conventional natural gas are given in [95, 96] (Table 4.8). Views on the available volumes of unconventional natural gas on the globe differ greatly. The data from [99] (Table 4.9) may be taken as initial data and the structure of their territorial distribution — from [2, 100].

Table 4.5. Oil resources (conventional and unconventional) by cost category in GEM-10R, million TJ

Cate-gory	Region										
	NA	EU	JK	AZ	SU	LA	ME	AF	CH	SA	World
1	357	248	0	17	718	731	3692	168	214	164	6309
2	361	97	0	13	571	374	714	143	197	80	2550
3	281	176	0	25	811	651	920	206	344	130	3544
4	668	244	0	29	983	794	2360	227	311	176	5792
5	319	55	0	155	139	109	937	59	97	29	1899
6	4150	340	0	1084	815	3843	1663	214	1772	214	14095
7	7258	601	0	1894	1428	6724	2911	374	3100	370	24660
8	12071	1613	0	2533	5275	11374	11718	1247	4985	1113	51929
Total	25465	3374	0	5750	10740	24600	24915	2638	11020	2276	110778

Table 4.6. Provision with oil resources, TJ/capita

Oil (category)	Region										
	NA	EU	JK	AZ	SU	LA	ME	AF	CH	SA	World
Very cheap (1)	1.3	0.5	0.0	0.8	2.5	1.63	13.6	0.3	0.2	0.1	1.2
Cheap (2, 3)	2.3	0.5	0.0	1.8	4.8	2.29	6.0	0.7	0.5	0.1	1.2
Expensive (4, 5)	3.6	0.5	0.0	8.8	3.9	2.02	12.2	0.6	0.3	0.1	1.5
Very expensive (6)	15.0	0.6	0.0	51.6	2.8	8.58	6.1	0.4	1.5	0.1	2.6
Very expensive unconventional (7, 8)	70.0	4.0	0.0	211	23.2	40.4	54.0	3.2	6.7	1.0	14.5
Total	92.3	6.1	0.0	274	37.2	54.9	91.9	5.3	9.2	1.5	20.9

Table 4.7. Oil production, consumption of oil and oil products and capacities of oil refineries by region (2000), million TJ/year

Region	Production	Consumption	Capacities of oil refineries
NA	20.1	41.2	38.4
EU	13.8	31.6	34.3
JK	0.0	14.9	15.4
AZ	1.5	1.9	1.9
SU	16.6	7.3	18.8
LA	21.9	12.7	16.7
ME	54.3	10.2	13.3
AF	8.1	3.4	6.2
CH	7.5	11.6	11.3
SA	7.0	12.3	15.0
World	150.8	147.1	171.5

Table 4.8. Resources of conventional natural gas, million TJ

Region	Resources			
	Discovered	Probable (50% of discovery probability)	Additionally possible for discovery	Total
NA	375	781	821	1977
EU	246	198	171	615
JK	0	0	0	0
AZ	42	104	7	153
SU	1980	1613	338	3931
LA	265	377	58	700
ME	1743	1249	77	3069
AF	160	350	8	518
CH	50	265	17	332
SA	274	368	63	705
World	5135	5305	1560	12000

Table 4.9 comprises rather speculative resources of methane hydrates. Their geological reserves are supposed to be huge: 30.3 billion TJ in permafrost (of which 23.5 billion TJ in Russia and 2.7 billion TJ in the USA) and about 720 billion TJ in water areas of (usually deep-water) seas [99, 101]. However, the extractable reserves of methane hydrates are tens and hundreds times less (we used the coefficient 0.01). Their production cost will be very high (probably above US\$ 10–15/GJ) due to

natural conditions of their occurrence and the complex technology of their production.

Table 4.9. Resources of unconventional natural gas

Gas source	World geological resources, billion TJ	Technically recoverable resources, million TJ		
		North America with production cost (US$/GJ)		Other countries
		to 7.5	to 3	
Dense sandstones	2.7–3.6	180	54	220–400
Oil shales	25.0–26.5	36	10	50–250
Coal beds	2.2–2.3	360	7	140–220
Aqueous solutions of zones with abnormal geopressure	1.1–6.1	160	0	360–860
Gas hydrates in permafrost areas	30.3	27	0	235
Gas hydrates in water areas of seas	720			
T o t a l	781–789	763	71	1005–1969

Based on these data and the work [102], we made an assessment of natural gas resources by region of the world in 1994 [89].

In the estimates obtained later by H.-H.Rogner [48, 93] on natural gas resources the gas resources of more expensive cost categories, basically gas from unconventional sources, were increased considerably. In the last version of the GEM-10R model the estimates of natural gas resources were taken on the base of H.-H.Rogner's data (Table 4.10). According to these data 97% of resources refers to three most expensive categories (6–8), of which 94% accounts for the most expensive eighth category.

The following regions are richest in natural gas at an acceptable extraction cost (categories 1–5) (TJ/capita): AZ — 35.8, SU — 28.4, ME — 19.2 and NA — 13.0 with the world average of 4.6 (Table 4.11). The JK, NA and CH regions are poorest in gas resources (<1.3 TJ/capita).

Table 4.12 presents volumes of the current natural gas production and consumption [98]. At such production volumes the natural gas resources of the first five categories will be enough for 270 years (this period is twice as long as that for oil).

Coal. Data on coal reserves are usually more reliable than those on oil and gas even for poorly explored deposits [91]. This fact is explained by specific features in formation of coal-bearing provinces. At the same time coals are characterised by a wide range of distinctions in their properties, in particular their combustion heat.

This circumstance being considered, the studies on resources become more complicated and laborious.

Table 4.10. Resources of natural gas (conventional and unconventional) by cost category according to the GEM-10R model, million TJ

Category	NA	EU	JK	AZ	SU	LA	ME	AF	CH	SA	World
1	496	336	0	88	1642	319	2024	164	46	294	5409
2	601	235	0	21	1890	336	966	223	193	235	4700
3	655	353	0	34	2730	580	1130	353	298	319	6452
4	353	147	0	21	848	164	525	92	67	113	2330
5	1470	210	0	588	1092	546	546	168	882	168	5670
6	2940	462	2	1258	1890	1260	1218	378	1008	420	10836
7	4410	672	4	1886	2856	1848	1848	588	1512	588	16212
8	256200	32172	6	63960	176736	191982	8526	16086	18144	24066	787878
Total	267125	34587	12	67856	189684	197035	16783	18052	22150	26203	839487

Table 4.11. Provision with natural gas resources, TJ/capita

Gas (category)	Region										
	NA	EU	JK	AZ	SU	LA	ME	AF	CH	SA	World
Very cheap (1)	1.8	0.6	0.0	4.2	5.7	0.7	7.5	0.3	0.0	0.2	1.0
Cheap (2, 3)	4.6	1.1	0.0	2.6	16.0	2.0	7.7	1.2	0.4	0.4	2.1
Expensive (4, 5)	6.6	0.6	0.0	29.0	6.7	1.6	4.0	0.5	0.8	0.2	1.5
Very expensive (6)	10.6	0.8	0.0	59.9	6.5	2.8	4.5	0.8	0.8	0.3	2.1
Very expensive unconventional (7, 8)	944	59.3	0.1	3136	621	433	38.3	33.2	16.3	15.8	152
Total	968	62.4	0.1	3231	656	440	61.9	36.0	18.4	16.8	159

Tables 4.13–4.15 present estimates of total economic coal resources (both discovered and additionally possible for discovery) expressed in energy units with regard to coal grade by the data of WEC [96] and other sources. Based on these data we estimated coal resources by cost category for the regions [89].

Table 4.12. Natural gas production and consumption by region (2000), million TJ/year

Region	Production	Consumption
NA	27.3	27.7
EU	10.9	17.3
JK	0.0	3.7
AZ	1.2	1.0
SU	25.5	20.7
LA	5.0	4.9
ME	12.2	8.7
AF	0.6	0.6
CH	1.1	1.3
SA	7.8	5.0
World	91.6	90.9

Table 4.13. Coal resources, million TJ

Region	Resources		Total
	Discovered	Additional	
NA	5760	13090	18850
EU	2770	7250	10020
JK	21	2	23
AZ	1780	8920	10700
SU	5720	49250	54970
LA	260	370	630
ME	10	5	15
AF	1960	1690	3650
CH	2640	3340	5980
SA	2160	1060	3220
World	23081	84977	108058

Table 4.14. Qualitative composition of discovered coal resources, million TJ

Region	Hard coals	Brown coals	Lignites	Total
NA	3256	1983	522	5761
EU	1745	74	949	2768
JK	21	0	0	21
AZ	1252	71	461	1784
SU	2600	2222	900	5722
LA	183	78	1	262
ME	9	1	0	10
AF	1940	18	0	1958
CH	1628	747	262	2637
SA	1275	239	650	2164
World	13909	5433	3745	23087

Table 4.15. Qualitative composition of additional coal resources, million TJ

Region	Hard coals	Brown coals	Lignites	T o t a l
NA	6929	3151	3010	13090
EU	6960	48	239	7247
JK	2	0	0	2
AZ	6900	2	2015	8917
SU	26250	15100	7900	49250
LA	235	127	4	366
ME	4	1	0	5
AF	1678	9	9	1696
CH	1776	1075	486	3337
SA	1011	17	35	1063
World	51745	19530	13698	84973

The estimates of coal resources taken in the last version of the GEM-10R model (Table 4.16) were obtained on the base of H.-H.Rogner's data [48, 93]. According to these data 63% of coal resources belongs to the most expensive category.

Table 4.16. Coal resources by cost category in GEM-10R model, million TJ

Cate-gory	Region										
	NA	EU	JK	AZ	SU	LA	ME	AF	CH	SA	World
1	1915	718	0	365	1386	76	0	466	605	126	5657
2	4469	1676	0	853	3234	176	0	1088	1411	294	13201
3	0	1260	0	7140	0	42	0	0	2856	1176	12474
4	2022	945	0	359	454	57	0	699	5594	378	10508
5	2472	1155	0	439	554	69	0	855	6838	462	12844
6	2363	1058	5	884	10395	132	57	302	3345	151	18692
7	2888	1294	6	1080	12705	162	69	370	4089	185	22848
8	20916	9492	13	7883	92442	1218	504	2688	29694	1344	166194
Total	37045	17598	24	19003	121170	1932	630	6468	54432	4116	262418

The following regions are richest in coal at an acceptable extraction cost (categories 1–5) (TJ/capita): AZ — 436.0, NA — 39.4, SU — 19.5 and CH — 14.4 at the world average value of 10.4 (Table 4.17). The JK, ME, LA and SA regions are least provided with coal (below 1.6 TJ/capita).

Table 4.17. Provision with coal resources, TJ/capita

Coal	Region										
(category)	NA	EU	JK	AZ	SU	LA	ME	AF	CH	SA	World
Very cheap (1)	6.9	1.3	0.0	17.4	4.8	0.2	0.0	0.9	0.5	0.1	1.1
Cheap (2, 3)	16.2	5.3	0.0	380.6	11.2	0.5	0.0	2.2	3.5	0.9	4.9
Expensive (4,5)	16.3	3.8	0.0	38.0	3.5	0.3	0.0	3.1	10.3	0.5	4.4
Very expensive (6)	8.6	1.9	0.0	42.1	36.0	0.3	0.2	0.6	2.8	0.1	3.5
Most expensive (7, 8)	86.3	19.5	0.1	426.8	363.8	3.1	2.1	6.1	28.0	1.0	35.7
T o t a l	134.2	31.8	0.1	904.9	419.3	4.3	2.3	12.9	45.2	2.7	49.6

Table 4.18 presents information on the current coal production and consumption [96, 98, 103 – 105]. At such production rates the coal resources of the first five categories will be enough for 600 years.

Table 4.18. Coal production and consumption (2000), million TJ/year

Region	Production	Consumption
NA	25.5	24.9
EU	10.2	14.6
JK	0.2	6.0
AZ	6.6	2.0
SU	8.3	7.3
LA	1.8	1.1
ME	0.0	0.4
AF	5.2	3.7
CH	20.9	21.5
SA	11.1	10.3
World	89.8	91.8

4.3. Nuclear energy resources

Uranium. The unique character of uranium as an energy source is determined by dependence of the resource volume not only on its content in the field, but on the technology of use as well. In Tables 4.19, 4.20 economic uranium resources are given for the open and closed cycles, respectively [89]. These tables are obtained basically from the data in [96, 106]. Uranium resources are divided into discovered and additionally possible for discovery.

In GEM-10R the use is made of estimates of uranium resources obtained from Tables 4.19, 4.20, the data from [93] and other sources. The cost characteristics of nuclear fuel resources for thermal and fast reactors (uranium-235 and uranium-238) are presented in Table 4.21 and the amount of resources — in Tables 4.22 and 4.23.

The following regions are richest in uranium-235 of all cost categories (1–8) (TJ/capita): AZ — 76.0, NA — 5.2 and SU — 3.9 at the world average of 1.4 (Table 4.24). The SA, ME, JK and EU regions are least provided with uranium-235 (<0.3 TJ/capita).

Table 4.19. Uranium resources for use in the open-cycle, million TJ

| Region | Resources | | | | | |
| | Discovered | | Additional | | T o t a l | |
	1	2	1	2	1	2
NA	83	108	258	212	341	320
EU	20	36	267	35	287	71
JK	0	13	0	2	0	15
AZ	158	20	86	36	244	56
SU	156	74	269	386	425	460
LA	28	10	32	7	60	17
ME	9	0	13	20	22	20
AF	175	47	146	30	321	77
CH	7	10	156	168	163	178
SA	10	13	0	8	10	21
World	646	331	1227	904	1873	1235

N o t e. Production cost of cheap (1) uranium is up to US$ 80/kg, that of expensive (2) — US$ 80–130/kg.

Table 4.20. Uranium resources for use in the closed cycle, billion TJ

| Region | Resources | | | | | |
| | Discovered | | Additional | | T o t a l | |
	1	2	1	2	1	2
NA	5.2	6.8	16.1	13.3	21.3	20.0
EU	1.3	2.3	16.7	2.2	17.9	4.4
JK	0.0	0.8	0.0	0.1	0.0	0.9
AZ	9.9	1.3	5.4	2.3	15.3	3.5
SU	9.8	4.6	16.8	24.1	26.6	28.8
LA	1.8	0.6	2.0	0.4	3.8	1.1
ME	0.6	0.0	0.8	1.3	1.4	1.3
AF	10.9	2.9	9.1	1.9	20.1	4.8
CH	0.4	0.6	9.8	10.5	10.2	11.1
SA	0.6	0.8	0.0	0.5	0.6	1.3
World	40.4	20.7	76.7	56.5	117.1	77.2

N o t e. Production cost of cheap (1) uranium is up to US$ 80/kg, that of expensive (2) — US$ 80–130/kg.

Table 4.21. Cost of uranium-235 and uranium-238, US$/GJ

Resource	Category							
	1	2	3	4	5	6	7	8
Uranium-235	0.50	0.60	0.73	0.88	1.09	1.32	1.55	1.78
Uranium-238	0.53	0.63	0.77	1.06	1.63	–	–	–

Table 4.22. Uranium-235 resources by cost category (for use in the open cycle) according to GEM-10R model, million TJ

Cate-gory	Region										
	NA	EU	JK	AZ	SU	LA	ME	AF	CH	SA	World
1	0.0	1.7	0.0	0.0	195.4	17.5	0.0	109.5	6.7	0.0	330.8
2	480.1	30.5	0.0	254.7	269.5	116.3	9.0	112.0	21.2	0.0	1293.3
3	786.2	86.0	2.2	46.0	492.8	260.1	8.7	88.8	505.1	33.5	2309.4
4	23.2	6.6	2.2	194.3	23.4	8.7	0.8	56.1	168.4	3.3	487.0
5	27.1	7.7	2.6	226.7	27.3	10.2	0.9	65.5	196.5	3.9	568.4
6	31.0	8.8	3.0	259.1	31.2	11.6	1.0	74.9	224.6	4.4	649.6
7	34.9	9.9	3.4	291.5	35.2	13.1	1.1	84.2	252.7	5.0	731.0
8	38.7	11.0	3.7	323.9	39.1	14.5	1.3	93.6	280.7	5.5	812.0
Total	1421.2	162.2	17.1	1596.2	1113.9	452.0	22.8	684.6	1655.9	55.6	7181.5

Table 4.23. Uranium-238 resources by cost category (for use in the closed cycle) according to GEM-10R model, billion TJ

Cate-gory	Region										
	NA	EU	JK	AZ	SU	LA	ME	AF	CH	SA	World
1	0	105	0	0	12214	1092	0	6846	420	0	20677
2	30009	1907	0	15918	16842	7266	561	6999	1325	0	80827
3	49140	5374	139	2877	30801	16254	546	5550	31569	2094	144344
4	4356	1238	433	36422	4394	1635	142	10527	31582	624	91353
5	5325	1513	504	44541	5371	1998	173	12867	38600	762	111654
Total	88830	10137	1076	99758	69622	28245	1422	42789	103496	3480	448855

Table 4.24. Provision with uranium-235 resources, TJ/capita

Uranium	Region										
(category)	NA	EU	JK	AZ	SU	LA	ME	AF	CH	SA	World
Very cheap (1)	0.0	0.0	0.0	0.0	0.7	0.0	0.0	0.2	0.0	0.0	0.1
Cheap (2, 3)	4.6	0.2	0.0	14.3	2.6	0.8	0.1	0.4	0.4	0.0	0.7
Expensive (4, 5)	0.2	0.0	0.0	20.0	0.2	0.0	0.0	0.2	0.3	0.0	0.2
Very expensive (6)	0.1	0.0	0.0	12.3	0.1	0.0	0.0	0.2	0.2	0.0	0.1
Most expensive (7, 8)	0.3	0.0	0.0	29.3	0.3	0.1	0.0	0.4	0.4	0.0	0.3
T o t a l	5.2	0.3	0.1	76.0	3.9	1.0	0.1	1.4	1.4	0.0	1.4

Table 4.25 presents the current level of nuclear energy development [96, 98].

Thorium. In the nearest decades there are no plans to develop thorium-based nuclear energy because of technical and mainly economic reasons. Nonetheless, thorium seems to be a potentially important energy resource. Therefore, Table 4.26 presents estimates of its economic resources that are obtained from the data in [107, etc].

Table 4.25. Nuclear energy consumption (2000), capacities of nuclear power plants in operation, under construction and planned and nuclear power generation (1998)

Region	Nuclear energy consumption, million TJ/year	Nuclear capacities, GW		Nuclear power production, TWh/year
		In operation	Under construction	
NA	9.4	108.1	0	741.2
EU	10.6	135.7	4.2	886.3
JK	4.6	56.4	7.5	392.1
AZ	0.0	0.0	0.0	0.0
SU	2.4	36.2	8.0	179.8
LA	0.2	2.9	1.7	19.0
ME	0.0	0.0	0.9	0.0
AF	0.1	1.8	0.0.0	13.6
CH	0.6	7.2	9.3	48.9
SA	0.2	1.8	1.3	10.5
World	28.1	350.1	32.9	2291.4

Table 4.26. Thorium resources, billion TJ

| Region | Resources | | | | | |
| | Discovered | | Additional | | T o t a l | |
	1	2	1	2	1	2
NA	1.5–2.9	0.5	8.2–11.2	0–0.2	9.7–11.7	0.5–0.7
EU	3.2–9.9	3.1–9.8	0.2	12.3	3.4–11.1	15.4–22.1
JK	0	0	0–6.0	0–6.0	0–6.0	0–6.0
AZ	0.4	0.1	0	0.2	0.4	0.3
SU	1.6	–	–	1.6	1.6	1.6
LA	3.6	0	20.0	6.0	23.6	6.0
ME	0	0.3–0.4	0.6	5.6–6.4	0.6	5.9–6.8
AF	0.2–0.4	0.3–0.4	0.2–0.4	0.2	0.4–0.8	0.5–0.6
CH	–	–	–	–	–	–
SA	0–6.4	0–6.4	–	–	0–6.4	0–6.4
World	11–25	4–18	29–38	26–33	40–63	30–51

N o t e. Production cost of cheap (1) thorium is up to US$ 80/kg, that of expensive (2) one — US$ 80–130/kg, "–" — no data.

4.4. Potential of renewable energy sources

Biomass. Annually about 250 billion t of bioproducts (or 4.5 billion TJ) on a dry basis are formed on the Earth [108]. Some 0.5% is consumed as food and 0.84% as fuel.

Primarily wood and waste of woodworking and agriculture and also domestic (urban) waste are used now for energy purposes (Table 4.27) [96, 109]. The contribution of energy biomass plantations is negligible as yet. Biomass is often called "oil of poor countries". Actually, based on WEC's estimates, its fraction in the fuel balance of 40 least developed countries exceeds 70% [96].

Table 4.27. Biomass consumption for energy purposes (1990), million TJ/year

| Region | Wood | | Products of biomass | Waste (agricultu- | Total |
	Total	Noncommercial	plantations	ral & urban)	
NA	2.60	0.25	0.08	0.25	2.93
EU	1.26	0.45	0.0	0.39	1.65
JK	0.05	0.01	0.0	0.03	0.08
AZ	0.11	0.02	0.0	0.05	0.16
SU	1.17	0.23	0.0	0.11	1.28
LA	7.35	5.29	0.34	0.63	8.32
ME	0.29	0.21	0.0	0.25	0.54
AF	7.39	7.18	0.0	0.42	7.81
CH	2.65	2.31	0.0	0.71	3.36
SA	10.05	9.50	0.0	1.34	11.39
World	32.92	25.45	0.42	4.18	37.52

Solar energy serves as an energy source for forming primary organic substances in nature. Its supply depends on geographical location and climatic peculiarities of the region and ranges, as a rule, from 4 to 8 GJ/m^2 per year. The volumes of annual biomass growth and efficiency of solar energy utilisation for different types of ecosystems are presented in Table 4.28.

The issue of biomass resources for the energy sector has long stirred specialists. Huge theoretical resources estimated in [2] at 1700 million TJ/year are inviting. However, not all of them are suitable for energy usage for diverse reasons. Rigid constraints of agriculture and industry of structural materials and the social sphere and ecology seem to be basic. Hence, according to [2] and [109] the technically realisable biomass resources for the energy sector will make up about 230–250 million TJ/year.

Table 4.28. Annual biomass growth and efficiency of solar energy utilisation for different types of ecosystems

Ecosystem	Annual biomass increase, MJ/(year·m^2)	Efficiency of solar energy utilisation, %
Pastures	50	0.9–1.2
Forests of temperate zones	130	2–3
Tropical forests	200	3–4
Ground bioplantations	380	5–7
Water bioplantations	1000	8–12

From the standpoint of other authors (in particular, [96]) this figure is too high. We also hold this viewpoint and, using the results of ([96, 110], etc.), our estimate of technical biomass resources is 96 million TJ/year (Table 4.29). Analysis has shown that deviations in estimates are explained basically by the approach to the idea of creating large "energy plantations". In [2], for example, the total realisable (i.e. economic) biomass potential of the Earth for energy purposes is taken equal to 150 million TJ/year, of which the fraction of biomass plantations is 59% (88 million TJ/year), that of wood and wood waste — 31% (47 million TJ/year), urban waste — 6% (9 million TJ/year) and agricultural waste — 4% (6 million TJ/year).

Bearing in mind an increasing food problem due to the fast growth of population in the world, it is hardly probable that vast areas of fertile soil will be allocated for energy biomass plantations. Expectations of sharp increase in productivity of crops and biomass plantations owing to application of large amounts of fertilisers are also not justified. A short-term effect will be substituted by sharp degradation of soils. The use of achievements of gene engineering is fraught with potential danger of uncontrolled organisms to emerge.

One should also not rely on an essential increase in the use of agricultural waste. Real threat of complete loss of stability for agricultural ecosystems because of humus depletion forces agrarians to forecast an inevitable transition of agriculture to a cyclic (closed) model of production in terms of an organic substance, which in essence deprives the energy sector of this resource base. To eliminate soil degradation and maintain a self-supporting humus balance, almost the same amount of organic substances as was planted is to be returned to fields annually [110].

One can also not neglect economic limitations on biomass fuel resources. A large portion of this fuel is scattered on land and its concentration requires great expenditures. Besides, the volumetric energy intensity of dry biomass is, as a rule, 3–4 times (and wet material is 5–7 times) lower than that of coal. This further increases labour input and cost of collection, delivery and use of biomass fuel.

Table 4.29. Technically possible biomass resources for energy purposes, million TJ/year

Region	Wood	Products of biomass plantations	Agricultural waste	Urban waste	Total
NA	8.99	0.58	2.56	0.24	12.37
EU	2.77	0.31	2.28	0.46	5.82
JK	0.71	0.00	0.29	0.13	1.13
AZ	0.21	0.10	0.40	0.02	0.73
SU	9.07	0.55	1.48	0.23	11.33
LA	20.33	0.66	2.60	0.10	23.69
ME	0.92	0.04	0.22	0.03	1.21
AF	11.38	0.23	1.02	0.04	12.67
CH	3.57	0.13	3.36	0.14	7.20
SA	13.86	0.34	5.33	0.16	19.69
World	71.81	2.94	19.54	1.55	95.84

Consideration of these factors allowed the estimation of economic biomass resources for energy purposes (Table 4.30). The table was constructed on the base of publications [96, 109–113]. The cost characteristics of biomass resources (US$/GJ) are: category 1 — 1.9, 2 — 2.6 and 3 — 3.5.

Resources of cheap biomass (category 1) in Table 4.30 are close to its current consumption in regions with the same production cost.

Table 4.30. Economic biomass resources for energy purposes, million TJ/year

Category	NA	EU	JK	AZ	SU	LA	ME	AF	CH	SA	World
1	2.9	1.6	0.1	0.2	2.7	8.5	0.6	7.9	3.5	11.4	39.4
2	4.6	1.4	0.6	0.2	5.3	10.5	0.4	3.7	0.6	3.6	30.9
3	4.0	2.0	0.3	0.2	3.0	4.0	0.2	1.0	3.0	4.0	21.7
Total	11.5	5.0	1.0	0.6	11.0	23.0	1.2	12.6	7.1	19.0	92.0

The following are the regions with the richest biomass resources of all cost categories (1–3), GJ/(year·capita): LA — 51.3, NA — 41.7, SU — 38.1, AZ — 28.6 and AF — 25.1 at an average world value of 17.4 (Table 4.31). The ME, JK and CH regions are poorest in biomass resources (<6 GJ/(year·capita).

Table 4.31. Provision with biomass resources, GJ/(year·capita)

Cost (category)	NA	EU	JK	AZ	SU	LA	ME	AF	CH	SA	World
Cheap (1)	10.5	2.9	0.5	7.6	9.3	19.0	2.0	15.7	2.9	7.3	7.4
Expensive (2)	16.7	2.5	3.6	11.4	18.3	23.4	1.3	7.4	0.5	2.3	5.8
Very expensive (3)	14.5	3.6	1.8	9.5	10.4	8.9	0.7	2.0	2.5	2.6	4.1
T o t a l	41.7	9.0	5.9	28.6	38.1	51.3	4.1	25.1	5.9	12.2	17.4

Hydroenergy. Table 4.32 presents estimates of hydroenergy resources by region of the world based on the analysis of a great number of publications involving appropriate information on each country [90]. The table contains theoretical, technically possible and economic resources as an average many-year electricity production per year (and installed capacity of hydropower plants).

Theoretical resources are understood as the total quantity of electricity that could be generated in this region, if the whole potential of natural water flows, with 100% efficiency, would be drawn down to sea level or the boundary with neighbouring regions. Routinely, input information for such estimates is based on meteorological data on atmospheric precipitation and topographic characteristics of the region. As a rule, the wettest areas have the highest hydroenergy potential.

Technically possible hydroenergy resources in Table 4.32 represent an annual amount of electricity that can be generated in the region by current and prospective technologies with regard to topographic and other natural limitations (i.e. by using all sites suitable for construction of hydropower plants), but neglecting social, economic and environmental constraints.

The economic (or commercial) resources represent a part of technically possible resources that can compete with alternative sources (as a rule, thermal power plants), after additional (social, environmental and juridical, etc.) constraints are taken into consideration.

Table 4.33 characterises utilisation level, of world hydroenergy resources. It shows that their economic potential has been already used by more than 50% in developed regions and by a mere 10–20% in developing regions. The other resources in developed regions are basically expensive, whereas developing regions possess a lot of cheap resources (Table 4.34).

Table 4.35, compiled from Table 4.34, presents estimates of the hydroenergy potential in energy units of the primary energy basis (the transformation ratio was taken equal to 0.85). Differentiated are the following categories of hydroenergy resources: 1) developed, 2) remained cheap, 3) remained expensive and 4) remained very expensive. Cheap resources include hydroenergy resources that ensure electricity generation at a cost that is no higher than the environmentally sound coal-fired power plants. For expensive and very expensive resources the electricity cost rises by a factor of 1.5 and more.

Table 4.32. World hydroenergy resources (installed capacity of hydropower plants and annual power production)

Region	Theoretical	Technically possible	Resources Total	Economic Utilised	Accessible to 2020
NA		295	230–244	165	25–76
	3120	1160	936–994	632	102–290
EU		433–449	300–308	152	36–65
	3017	1220–1285	862–891	623	92–172
JK		49	48	23	7–22
	806	152	149	98	21–47
AZ		28–31	22	11	2–3
	416	94–111	74	41	5–8
SU		690	330	65	48
	3942	2190	1095	229	169
LA		834–1117	434–601	119	71–122
	9529	3312–4443	1964–2460	582	309–492
ME		56	33	8	10
	193	155	99	22	24
AF		296–298	128	17	21–28
	3062	1192–1221	425	62	80–104
CH		670	265–301	77	69–107
	6117	2260	881–1000	222	286–369
SA		504–625	197–260	46	35–95
	8168	2119–2520	753–1128	177	132–401
World		3855–4280	1987–2275	693	324–576
	38370	13854–15469	7238–8315	2688	1220–2076

* Upper line — GW, lower line — TWh.

Table 4.33. Utilisation level of world hydroenergy resources, %

Region	Potential			Resources		
		Theore-tical	Technically possible	Total	Economic Utilised	Accessible to 2020
NA	1	100	37	30–32	18	3–9
	2		100	81–86	49	9–25
	3			100	57–61	10–29
EU	1	100	40–43	29–30	16	3–6
	2		100	69–71	36–38	7–14
	3			100	52–53	11–20
JK	1	100	19	18	12	2–6
	2		100	98	61	14–31
	3			100	62	14–32
AZ	1	100	23–27	18	9	1–2
	2		100	67–79	35–41	5–7
	3			100	46	7–11
SU	1	100	56	28	6	4
	2		100	50	10	8
	3			100	21	15
LA	1	100	35–47	21–26	4	3–5
	2		100	55–59	9–12	9–11
	3			100	17–21	16–20
ME	1	100	80	51	12	12
	2		100	64	14	15
	3			100	23	24
AF	1	100	39–40	14	2	3
	2		100	35–36	4	7–9
	3			100	11	19–25
CH	1	100	37	14–16	3	5–6
	2		100	39–44	8	13–16
	3			100	17–19	32–37
SA	1	100	26–31	9–14	1	2–5
	2		100	36–45	4–5	6–16
	3			100	10–15	18–36
World	1	100	36–40	19–22	6	3–5
	2		100	52–53	14–16	9–13
	3			100	26–30	17–25

N o t e: 1 — theoretical, 2 — technical, 3 — economic.

Table 4.34. Distribution of economic hydroenergy resources, TWh/year

Region	Total	Utilised resources (2000)	Remaining resources at average cost		
			c_α	$1,5\ c_\alpha$	$2.5\ c_\alpha$
NA	936–994	632	39–77	234–249	94–99
EU	862–891	623	65–84	216–223	86–89
JK	149	98	4	37	15
AZ	74	41	9	19	7
SU	1095	229	486	274	110
LA	1964–2460	582	866–1188	491–615	196–246
ME	99	22	42	25	10
AF	425	62	229	106	43
CH	881–1000	222	403–480	220–250	88–100
SA	753–1128	177	378–621	188–282	75–113
World	7238–8315	2688	2519–3219	1811–2080	725–833

N o t e. c_α is the cost of electricity generated by an alternative environmentally sound coal-fired power plant.

Table 4.35. Economic hydroenergy resources by cost category, million TJ/year

Category	Region										
	NA	EU	JK	AZ	SU	LA	ME	AF	CH	SA	World
1	2.41	2.10	0.39	0.17	0.96	1.74	0.09	0.20	0.72	0.47	9.25
2	0.24	0.31	0.02	0.03	2.05	4.26	0.18	0.96	1.87	2.00	11.92
3	1.03	0.94	0.16	0.09	1.16	2.41	0.11	0.46	0.99	1.13	8.48
4	0.41	0.37	0.06	0.03	0.46	0.94	0.04	0.18	0.40	0.40	3.29
Total	4.09	3.72	0.63	0.32	4.63	9.35	0.42	1.80	3.98	4.00	32.94

The volumes of economic hydroenergy resources in Table 4.35 are the average values of the corresponding ranges in Table 4.34. Nonetheless, they should be treated as rather optimistic. As seen from practice, hydroenergy resources tend toward decrease of their economic portion due to additional constraints and tightening of those taken into account earlier.

Almost 94% of still unused cheap resources is concentrated in five regions: the former USSR, Latin America, Africa, South and Southeast Asia and China. Their development will probably give rise to additional problems, primarily environmental and social, that will be due in particular to flooding of vast territories. In Brazil, for example, an increase of hydropower capacity by 40 GW will lead to flooding of 26 thousand km^2, which makes up 0.3% of the state area. The coefficient of specific flooding is 650 km^2/GW. The next 132 GW of hydropower, drawn from the remaining economic hydroenergy potential, will result in additional flooding of 144 thousand km^2 or 1.7% of the country's territory (the coefficient of specific flooding is 1100 km^2/GW) [114]. Thus, based on current estimates, development of the whole economic hydroenergy potential of Brazil will result in flooding of about 2% of the national territory, a large part of which belongs to agricultural areas, and the remainder is covered by forest stands. It is hard to believe that society will agree with this.

The hydropower industry of the former USSR, Latin America, Africa and China is characterised by essential remoteness of areas rich in water resources from power consumption centres. The distance is 1500–2000 km for Africa, 2500 km for Latin America and China and 4000–5000 km for the former USSR. Expenditures on installation of transmission lines of such a length will have a drastic influence on electricity cost for consumers and hence, on competition between hydropower and coal-fired power plants.

In South and Southeast Asia a high hydroenergy potential is concentrated in mountainous areas of the mainland and on islands of the Pacific, where there are often no significant consumers of electricity.

More than half the remaining cheap hydroenergy resources can be found in the tropical zone. The experience of operating hydropower plants there shows that creation of large water reservoirs in such areas inevitably results in a complex of heavy environmental and social (including medical) problems. Decay of algae and "blossoming" of stagnant water degrade its quality to such an extent that it becomes inadequate for drinking not only from the reservoir, but even downstream. In the tropical climate, water reservoirs prove to be a source of many diseases (malaria, etc.).

The mentioned circumstances and constraints being considered, some part of cheap resources will be shifted to a category of expensive ones and even beyond the limits of an economic class.

Technical and economic indices of hydropower plants that correspond to cost categories of Table 4.35 will be described in Chapter 5 (see Table 5.4).

Wind energy. About 25% of solar radiation reaching the lower layers of the atmosphere is converted to kinetic energy of wind [109]. However, only a small fraction finds application. Accessible resources are limited by the volume of surface atmosphere over dry land and coastal shoals with a height of 200 m in areas with an average annual wind velocity of above 5m/s (at a height of 10 m). Such a wind velocity corresponds to the third class of wind resources in accordance with the USA scale (Table 4.36).

Areas with the fifth class of wind and higher where the average annual specific wind velocity exceeds 6 m/s at a height of 10 m and 7.7 m/s at a height of 50 m (the specific power of wind flow exceeding 250 and 500 W/m^2 respectively) are considered at present as proper for wind energy development. Improvement of wind turbines may transfer areas with the fourth and then the third classes of wind to quite reasonable for wind energy utilisation.

Table 4.36. Classification of wind energy resources in terms of average annual wind velocity (or average annual specific wind power at heights of 10 and 50 m from the earth's surface)

Class	Height 10 m		Height 50 m	
	Wind velocity, m/s	Specific power, W/m^2	Wind velocity, m/s	Specific power, W/m^2
1	0–4.4	0–100	0–5.6	0–200
2	4.4–5.1	100–150	5.6–6.4	200–300
3	5.1–5.6	150–200	6.4–7.0	300–400
4	5.6–6.0	200–250	7.0–7.5	400–500
5	6.0–6.4	250–300	7.5–8.0	500–600
6	6.4–7.0	300–400	8.0–8.8	600–800
7	7.0–9.4	400–1000	8.8–11.9	800–1200

Meteorological data on the area of dry land and coastal zones with winds of the corresponding classes [109, 112] were applied for estimation of theoretical resources of kinetic energy of wind by region of the world (Table 4.37) [89]. For the world as a whole, wind resources total about 415 billion TJ/year, being more than twice as much as total extractable conventional resources of oil, gas and coal.

Table 4.37. Wind energy resources of class 3 and higher, billion TJ/year

Region	Class			Total
	5–7	4	3	
NA	63.4	24.8	28.2	116.4
EU	7.1	6.1	3.9	17.1
JK	0.2	0.3	0.9	1.4
AZ	10.4	5.7	9.4	25.5
SU	21.6	32.0	37.2	90.8
LA	18.0	12.1	15.5	45.6
ME	0.7	8.2	7.3	16.2
AF	3.2	39.4	34.4	77.0
CH	1.7	2.9	7.9	12.5
SA	1.8	3.0	8.0	12.8
World	128.1	134.5	152.7	415.3

Figure 4.4 illustrates geographical distribution of wind energy resources in terms of the average annual wind velocity [115]. As seen from the figure, the main wind potential is typical of the coastal zones. As a rule, these are the areas with high population density, developed communications, industry and agriculture. It turns out to be rather difficult to find suitable sites for "wind farms", if the environmental and social constraints due to specific features of wind turbines (noise, influence of telecommunications, hindrance to aviation, "visual pollution" of the environment, etc.) are taken into account. One should bear in mind that just the windmills themselves occupy only 1% of the "farm" area.

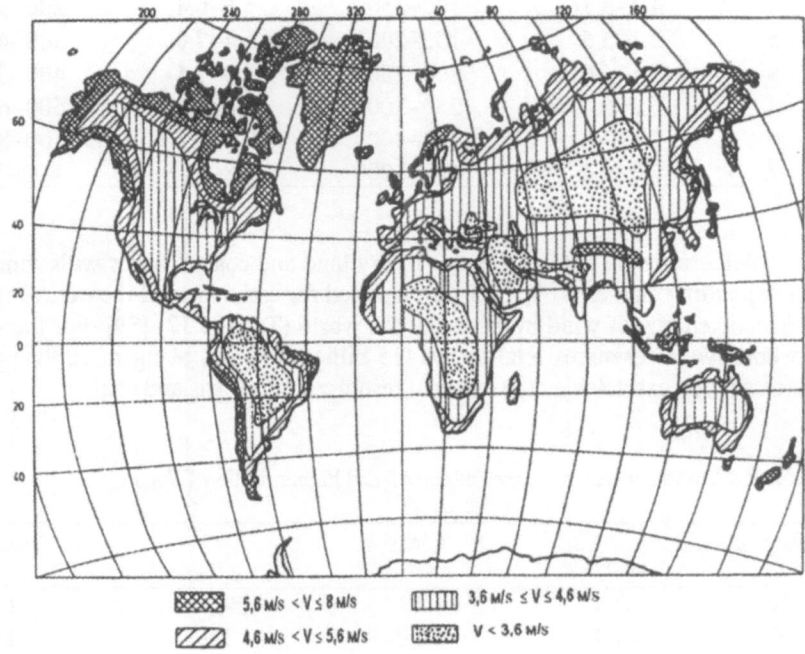

| ▨▨▨ | 5,6 M/s < V ≤ 8 M/s | ⅢⅢⅢ | 3,6 M/s ≤ V ≤ 4,6 M/s |
| ▨▨▨ | 4,6 M/s < V ≤ 5,6 M/s | ▨▨▨ | V < 3,6 M/s |

Figure 4.4. Geographical distribution of wind energy resources versus average annual wind velocity.

There is apprehension that large-scale use of wind turbines in the coastal zone can lead to essential climatic changes not only in this area, but deep in the mainland. They may cause, for example, changes in regimes of precipitation to such an extent that this will influence agriculture. The "interference" effect that implies mutual impact of wind turbines in a dense location considerably decreases accessible wind resources. The associated energy losses sharply rise at a distance between windmills of less than their 10 diameters (10 D). For a cluster of 100 windmills at a distance between them of 5 D, losses reach 40% [109].

Routinely, wind turbines are put into operation at a wind velocity above 3–5 m/s and to avoid failures they are removed from work at a wind velocity above 20–25 m/s. Since the wind turbine capacity has a cubic dependence on the wind velocity, elimination of high-energy storm winds from the resource base is noticeable, particularly in coastal areas.

A considerable wind potential is found on the coast of northern seas of Russia, Canada and Alaska. These areas are practically uninhabited.

A lot of works are devoted to comprehensive consideration of the mentioned factors when assessing a wind potential of different countries and regions. The results are summarised in [109, 112, 116] and formed the base for assessing potential, technically possible and economic resources of kinetic energy of wind for electricity generation (Table 4.38). Potential world resources proved to be equal to some 7 billion TJ/year, which does not exceed 2% of the theoretical value. Technically possible resources are by an order of magnitude lower. The economic resources in them make up some 20%. Estimates of the economic resources in Table 4.38 are thought to be optimistic.

Table 4.38. Wind energy resources for electricity production, million TJ/year

Region	Resources		
	Potential	Technically possible	Economic
NA	2000	202	75.2
EU	454	72	9.5
JK	23	4	1.0
AZ	432	40	9.5
SU	1520	150	42.0
LA	778	78	14.7
ME	268	28	1.1
AF	1267	130	6.4
CH	212	32	9.1
SA	220	33	7.7
World	7174	769	176.2

The economic wind resources are split into nine categories depending on the following parameters: 1) energy potential (wind class) and 2) distance from electricity consumption centres (Tables 4.39, 4.40).

Solar energy. The annual supply of solar energy to the earth's surface accounts for about 2500 billion TJ. The average annual intensity of solar radiation for the whole planet is approximately 160 W/m^2 [108]. However, it varies within sufficiently wide ranges depending on latitude and natural-climatic conditions: cloudiness, humidity, height above sea level, dustiness of the atmosphere, etc. In desert areas of Africa, Australia and America this value exceeds 250 W/m^2 and in high latitudes it does not reach 90 W/m^2.

Table 4.39. Qualitative characteristics of economic wind resources

Category	Wind potential	Distance to consumption centres
1		Close
2	High (wind class 5–7)	Far
3		Very far
4		Close
5	Medium (wind class 4)	Far
6		Very far
7		Close
8	Low (wind class 3)	Far
9		Very far

Table 4.40. Distribution of economic wind resources by category, million TJ/year

Category	Region										
	NA	EU	JK	AZ	SU	LA	ME	AF	CH	SA	World
1	3.0	1.0	0.2	1.0	0.5	1.2	0.1	0.2	0.6	1.2	9.0
2	1.0	0.5	0.3	0.5	0.5	2.5	0.1	1.0	1.0	1.2	8.6
3	26.2	2.0	0.0	3.5	11.0	5.0	0.2	0.2	3.0	2.3	53.4
4	6.0	1.1	0.1	0.6	3.0	0.7	0.0	0.6	0.6	0.5	13.2
5	2.5	0.2	0.0	0.7	1.0	1.0	0.2	0.7	0.3	0.2	6.8
6	5.0	0.5	0.0	0.0	5.0	0.0	0.0	0.2	0.4	0.3	11.4
7	14.0	2.5	0.4	1.4	7.0	1.8	0.2	1.4	1.4	1.0	31.1
8	5.5	1.8	0.0	1.8	6.0	2.0	0.3	1.9	0.7	0.3	20.3
9	12.0	0.0	0.0	0.0	8.0	0.5	0.0	0.2	1.1	0.7	22.5
Total	75.2	9.5	1.0	9.5	42.0	14.7	1.1	6.4	9.1	7.7	176.2

Figure 4.5 borrowed from [115] shows the supply of solar energy in terms of weather conditions.

Table 4.41 presents our estimates of economic solar energy resources for the energy sector [89]. Three categories of resources are separated based on the quantity of average annual solar energy supply:

— cheap (2200 kWh/m^2 or 8 GJ/m^2 per year and higher);
— expensive (on the average 1650 kWh/m^2 or 6 GJ/m^2 per year);
— very expensive (1100 kWh/m^2 or 4 GJ/m^2 per year).

Areas with an average annual intensity of solar radiation less than 120 W/m^2 were considered unpromising for a scaled solar energy development [89].

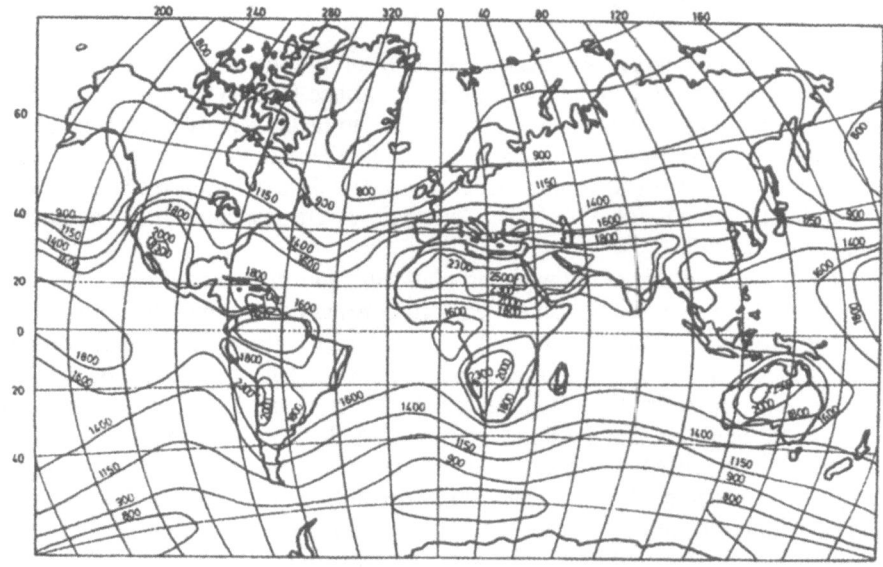

Figure 4.5. Geographical distribution of solar energy resources in terms of its annual supply, kWh/(year·m^2).

Table 4.41. Solar energy resources, million TJ/year

Region	Category			Total
	Cheap	Expensive	Very expensive	
NA	8.0	0.6	1.8	10.4
EU	0.0	0.0	0.6	0.6
JK	0.0	0.0	0.2	0.2
AZ	24.0	2.7	0.4	27.1
SU	0.0	1.8	3.0	4.8
LA	8.0	3.0	0.5	11.5
ME	72.0	3.0	0.0	75.0
AF	40.0	6.0	0.0	46.0
CH	0.0	0.3	4.0	4.3
SA	0.0	2.4	1.2	3.6
World	152.0	19.8	11.6	183.5

The calculations were made on the assumption that the following territories in appropriate "solar" zones will be used for solar energy utilisation: 0.2% in the areas with cheap solar energy (usually deserts), 0.1% in the areas with expensive solar energy (as a rule, semideserts), 0.01% in the areas with very expensive solar energy (these are primarily major agricultural areas and forest stands).

The technical and economic indices of solar power plants for three categories of resources that are presented in Table 4.41 will be analysed in Chapter 5 (Table 5.6).

Geothermal energy. An average density of thermal flow from the Earth's interior accounts for 0.060–0.065 W/m^2 at its surface [109], which is 2–4 thousand times less than that of solar radiation flow. Due to heterogeneity of thermophysical properties of the mantle and the Earth crust, the thermal flow density in different areas varies through a wide range. In some places it does not even reach 0.03 W/m^2, in the others exceeds 0.5 W/m^2. The latter are of greatest interest for geothermal energy development.

Fault zones faults of lithospheric platforms and places with high geological activity are, as a rule, the richest in geothermal resources. Among them are: Pacific Fire Ring (embracing the Pacific coast of North and Latin America, New Zealand, some islands of Oceania and Southeast Asia, Japan, the Kuril Islands and Kamchatka), the Alps- Himalayas mountain chain, Central Asia, East Africa, the Red sea area and the Central-Atlantic reef reaching Iceland (see Figure 4.6).

By EPRI's (USA) estimates the theoretical resources of geothermal energy on the Earth to a depth of 3 km equal approximately 41,000 billion TJ/year (Table 4.42). Within a depth of 5 km the theoretical resources are 3–4 times larger (about 140,000 billion TJ) [109]. These are tentative figures, but they characterise geothermal resources as practically inexhaustible.

Table 4.42. Potential geothermal energy resources, billion TJ/year [117]

Region	Temperature level, °C				Total
	To 100	100–150	150–250	> 250	
North America	7030	1200	322	21	8573
Europe	1500	60	18	3	1581
USSR	6880	260	62	6	7208
Latin America	4840	910	264	6	6020
Asia	7470	790	208	12	8480
Africa	5130	320	70	9	5529
Australia and Oceania	3220	260	170	9	3659
World	36070	3800	1114	66	41050

Geothermal energy resources, however, can not be considered as renewable ones in full measure, as far as they are represented basically by heat accumulated by rock and water reservoirs over many million years. The heat being extracted quickly, the rock will cool off (since its heat conductivity is relatively low) and the necessity will arise to develop new areas of accumulated heat.

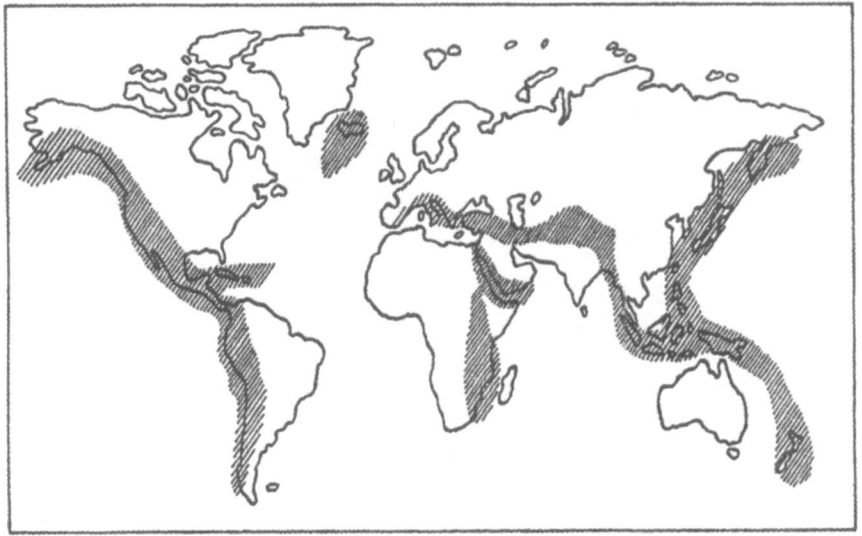

Figure 4.6. Geographical distribution of geothermal energy resources [109]

The most important characteristic of geothermal resources is the temperature at which energy is extracted. Table 4.42 shows that more than 97% of theoretical resources of geothermal energy belongs to the class of low-grade resources (88% with the temperature below 100°C and 9% with the temperature $100-150^{\circ}$C). In essence they are unsuitable for electricity generation. Resources of a higher grade account for less than 3% (2.7% with the temperature $150-250^{\circ}$C and solely 0.16% with the temperature above 350°C). However, the temperature level of even these resources is very low, being the main reason for low energy effectiveness of geothermal power plants. Their efficiency is, as a rule, 10–20%.

Geothermal resources may be split into four groups:

1) conventional hydrothermal (underground reserves of hot water and steam);

2) unconventional hydrothermal (reservoirs of hot water and steam under abnormally high pressure at a large depth);

3) petrothermal (heat accumulated by dry rock);

4) thermal energy of magma.

Resources of group 1 have long and successfully been in use. They are much cheaper than the others. Their fraction in total resources, however, is very negligible (for example, about 2% in the USA). Besides, they are primarily of low grade. In essence there are no fundamental technical limitations on using unconventional hydrothermal resources (resources of group 2), but they prove to be very expensive.

The technology of heat extraction from dry rock (group 3 of resources) is still at the R&D stage. Now it is evident that this technology will also be very expensive. The largest portion of expenditures is due to drilling and preparation of deep wells for exploitation, each of them being a rather sophisticated engineering construction.

According to [118] development of the thermal resources of hot dry rock is advisable solely in areas with an abnormally high geothermal gradient that exceeds $50-70^{\circ}C/km$ deep down (at an average value for "normal" areas of $25-35^{\circ}C/km$). From experimental results in Fenton Hill (USA, 1986) at the gradient $50-70\ ^{\circ}C/km$ one well 3.6 km deep (the rock temperature is $240^{\circ}C$) provides a 1 MW power unit with heat. For effective well exploitation over a long time span each should have an area of $0.1-1.0\ km^2$ (possibly more). Hence, a 1000 MW geothermal power plant will require a territory of $100-1000\ km^2$. Specific power of the geothermal power plant will account for $1-10\ MW/km^2$. For comparison, the peak specific power of the solar power plant is $50-100\ MW/km^2$, and the average annual specific power for different climatic zones ranges from $5-15$ to $10-25\ MW/km^2$ at an efficiency of solar radiation conversion to electric power of 10 and 20%, respectively.

Resources with a higher thermal gradient ($80-90^{\circ}C/km$) could offer better economic characteristics. Such places on Earth, however, are few in number.

Resources of group 4 that are connected with magma have the highest temperature. But their development is expected in a very remote future and they will obviously turn out to be even more expensive.

Intensive development of geothermal resources, especially heat of rock, in a limited area can trigger earthquakes (as a result of hydraulic faults of rock and increasing heterogeneity of its properties) and soil sag on large territories. Utilisation of geothermal energy creates other environmental and social problems due to extraction of large volumes of diverse harmful gases and brines.

Table 4.43 presents estimates of economic geothermal energy resources by region that were prepared by us for the GEM-10R model on the base of the data from [109, 112, 119] and others. Conventional geothermal energy resources (resources of group 1) are considered to be cheap. A class of expensive resources comprises geothermal energy of group 2 (unconventional hydrothermal resources), and geothermal resources of groups 3 and 4 (heat of dry rock and magma) are a class of very expensive ones.

Table 4.44 illustrates the current volumes of geothermal energy consumption for electricity and heat production [96, 103, 105].

Table 4.43. Economic resources of geothermal energy, million TJ/year

Region	Geothermal energy			Total
	Cheap	Expensive	Very expensive	
NA	2.5	6.5	10.0	19.0
EU	1.5	0.5	0.9	2.9
JK	0.7	1.4	0.8	2.9
AZ	0.4	0.3	0.4	1.1
SU	0.9	2.0	1.5	4.4
LA	2.8	8.0	9.0	19.8
ME	0.3	1.0	0.6	1.9
AF	2.0	5.5	5.0	12.5
CH	0.8	1.0	0.8	2.6
SA	2.5	7.5	7.0	17.0
World	14.4	33.7	36.0	84.1

Table 4.44. Use of geothermal energy for electricity and heat production (1990)

Region	Electricity		Heat	
	MW	GWh	MW	Thousand TJ
NA	2842	16900	465	1.4
EU	574	3600	3990	64.4
JK	270	1360	3220	24.5
AZ	264	2070	270	6.5
SU	11	25	1130	15.0
LA	870	5520	23	0.2
ME	0,4	1	103	1.4
AF	45	350	38	0.4
CH	28	93	2150	7.0
SA	1031	7500	1800	20.2
World	5935	37419	13189	141.0

Chapter 5

TECHNOLOGIES OF ENERGY CONVERSION AND FINAL CONSUMPTION

5.1. General remarks

This chapter addresses the forecasted technical, economic and environmental indices of technologies for production of final energy forms (electrical, thermal, mechanical and chemical) and for production of intermediate energy carriers (motor fuels, residual fuel oil, substitute gas, methanol and hydrogen) that are presented in the GEM-10R model. The authors mean here the generalized technologies (macrotechnologies) for production of any energy form or carrier using one or another primary energy resource (or energy carrier).

The given data are largely the result of forecast studies on energy technologies for the long-term future, performed in 1984–1994 [62]. The key elements of the research methodology are 1) thermodynamic analysis of technological processes that underlie the technology, 2) technical and economic analysis of the installation based on the technology. The studies involved thermodynamic models of extreme intermediate states and aggregated models of the corresponding technological installations based on estimations of changing relative weight and size characteristics of their elements [62].

The technological installations in the GEM-10R model are characterised by average indices for the region and for the operation period. The reason is that both new and old installations of the given type have different characteristics and operate in the region at every instant. Furthermore, the indices can substantially differ depending on an operation mode. The numerical values of the economic indices are given in 1990 US $ (disregarding inflation).

The specific reduced costs for each type of technological installations were calculated by the formula

$$C_{\text{mod}} = (a_k \text{K} + S)\frac{\eta\gamma}{h},$$

where

$$a_k = p/[1 - (1 + p)^{-T_{sl}}].$$

Here K — specific investments, US$/kW; S — specific constant operating costs, US$/(kW/year); p — a discount rate, 1/year; T_{sl} — service life of technological installation, years, η — efficiency, h — annual number of utilisation hours, $\gamma = 278$ kW/h/GJ. In the studies the discount rate was taken equal to p=0.05 1/year for all the world regions and calculated periods.

It should be mentioned that the tables in the given chapter give two values of production costs of a corresponding energy carrier (or final energy form) for each technology:

1) per unit of generated energy with consideration for fuel constituent (for example for power plants this is the value of electric power cost C_{el}, cent/(kW/h));

2) per unit of power input without consideration for fuel constituent — C_{mod}, US$/GJ p.i.; this value is used in GEM-10R as a coefficient of the objective function (see Section 2.3).

In the general case, economic comparison of technologies based on the first type of values on a world energy system scale is not correct. The only purpose of its introduction in the tables is to afford the opportunity to quickly, though tentatively, estimate comparative efficiency of the considered technologies (with conditionally accepted fuel cost C_f US$/GJ). The indices of C_{mod} do not have such a property. This has two reasons: there is no fuel constituent in them and they were reduced to the unit of the supplied energy (therefore, for the technology with lower efficiency the value C_{mod} will be smaller). Let us point out that the model GEM-10R takes into account the so-called "fuel constituent" of costs automatically (by the balance equations).

The information presented in the following sections is based both on review of the published data and on the authors' works that are presented in more detail. In particular, a detailed description of wind power plants and Space power systems is given. The latter even form a separate section.

The economic indices of most of the technologies vary with time — improve towards the end of the 21st century owing to technological progress or deteriorate due to either stricter environmental requirements or unfavorable conditions. These changes were estimated in an expert way. Basically, they correspond to the anticipated trends of technological progress in energy including the trend of accelerated updating of technologies with growing scales of their application ("learning by doing") — see, for example [120].

5.2 Technologies of electric power generation

5.2.1. Fossil fuel- and hydrogen-fired power plants

The model GEM-10R offers the opportunity to model power plants on different forms of fossil fuel: oil products (fuel oil), natural gas or substitute natural gas, coal, biomass, methanol and hydrogen. Liquid and gaseous fuel-fired power plants can be both base and peak, whereas power plants on solid fuel (coal and biomass) can be only base ones.

The main anticipated trends of technological progress of fossil fuel- and hydrogen-fired thermal power plants are:

1) transition to the integrated-gasification combined cycle installations, increase of temperature and compression ratio in gaseous cycle, increase of temperature and pressure in the steam cycle in the first quarter of the 21st century;

2) large-scale development of technologies for direct conversion of chemical energy of liquid and gaseous fuels (methanol, natural gas, SNG and hydrogen) to electricity by the middle of the century.

As a result the electric power generation efficiency should increase with some rise in specific capital investments and operating expenditures. The latter is explained by the complicated technological scheme and application of new (more expensive) construction materials for operation at higher thermodynamic parameters.

Table 5.1 presents technical and economic characteristics of electric power generation technologies on the basis of fossil fuels and hydrogen taken for the calculations on the model GEM-10R. The service life of all power plants is assumed equal to 30 years.

Table 5.2 presents base values of specific emissions into the atmosphere of sulfur oxide (SO_2), nitrogen oxide (NO_x), particulates of fly ash and carbon dioxide (CO_2) by electric power plants. Flue gases are additionally cleaned from harmful substances when needed (to meet the environmental constraints). To do this GEM-10R offers a set of appropriate cleaning technologies with their technical and economic indices. Cleaning technologies are started and the degree of cleaning is determined automatically in the model GEM-10R in the process of optimisation.

5.2.2. Nuclear power plants (NPPs)

The model offers two types of nuclear power plants: NPPs with thermal neutron reactors that use uranium-235 as a fuel and NPPs with fast neutron reactors (breeders) that use uranium-238 (as an additive to the main fuel — plutonium). Nuclear power plants of both types are base.

The prospects for improvement of NPP energy indices (efficiency) were accepted as quite restrained (Table 5.3). When forecasting economic characteristics it was assumed that cheapness of NPP owing to innovations would be normalized by increasing cost due to measures on enhancement of their safety. Service life of all NPP was assumed to be 40 years.

5.2.3. Hydropower plants (HPPs)

The key elements that determine technical and economic indices of HPP are: 1) height of a dam (drop of water levels in upper and lower reaches, 2) volumetric turbine water discharge 3) number of utilisation hours of the installed capacity and 4) characteristic of the site (mining and geographical conditions of construction). By virtue of these factors, technical and economic characteristics of HPP vary in the world regions which are taken into account in the model.

The energy efficiency indices of HPP have already reached the point of saturation. Their further improvement is practically impossible. Capital investments in HPP are supposed to increase steadily due to development of more expensive sites.

Table 5.1. Technical and economic indices of fossil fuel- and hydrogen-fired power plants

Thermal power plants	Year	Effici- ency (net)	h, h/year	K, US$/kW	S, US$/year / kW	C_f, US$/GJ	C_{el}, Cent/kW	C_{mod}, US$/ GJ (p.i.)
Fuel oil-fired, base	2025	0.43	6000	1100	39	2	3.61	2.32
	2050	0.48	6000	1150	41	3	4.28	2.71
	2075	0.52	6000	1180	42	4	4.85	3.01
	2100	0.54	6000	1200	42	5	5.44	3.16
Fuel oil-fired, peak	2025	0.35	1000	620	21	2	8.38	6.15
	2050	0.36	1000	625	21	3	9.36	6.36
	2075	0.38	1000	630	21	4	10.18	6.74
	2100	0.42	1000	640	21	5	10.74	7.53
Gas-fired, base	2025	0.45	6000	1050	37	2.3	3.67	2.29
	2050	0.50	6000	1100	40	3.3	4.31	2.69
	2075	0.53	6000	1130	41	4.3	4.91	2.93
	2100	0.55	6000	1150	41	5.3	5.48	3.07
Gas-fired, peak	2025	0.35	1000	600	20	2.3	8.45	5.92
	2050	0.36	1000	605	20	3.3	9.42	6.12
	2075	0.38	1000	610	20	4.3	10.23	6.49
	2100	0.42	1000	620	20-	5.3	10.76	7.26
Coal-fired	2025	0.38	6500	1450	58	1.5	3.91	2.63
	2050	0.43	6500	1460	58	2.0	4.18	2.99
	2075	0.48	6500	1480	59	2.4	4.34	2.39
	2100	0.50	6500	1500	60	2.7	4.52	3.58
Biomass-fired	2025	0.36	6000	1500	70	1.9	4.90	3.00
	2050	0.40	6000	1550	74	2.3	5.20	3.48
	2075	0.44	6000	1580	76	2.6	5.33	3.92
	2100	0.46	6000	1600	77	3.5	5.98	4.15
Methanol-fired, base	2025	0.45	6000	1050	37	9	9.03	2.29
	2050	0.50	6000	1100	40	10	9.14	2.69
	2075	0.53	6000	1130	41	11	9.46	2.93
	2100	0.55	6000	1150	41	12	9.87	3.07
Methanol-fired, peak	2025	0.35	1000	600	20	9	15.34	5.92
	2050	0.36	1000	605	20	10	16.12	6.12
	2075	0.38	1000	610	20	11	16.57	6.49
	2100	0.42	1000	620	20	12	16.51	7.26
Hydrogen-fired, base	2025	0.45	6000	1050	37	10	9.83	2.29
	2050	0.50	6000	1100	40	11	9.86	2.69
	2075	0.55	6000	1130	41	12	9.84	3.04
	2100	0.60	6000	1150	41	13	9.81	3.35
Hydrogen-fired, peak	2025	0.35	1000	600	20	10	16.37	5.92
	2050	0.40	1000	605	20	11	16.02	6.80
	2075	0.45	1000	610	20	12	15.75	7.69
	2100	0.50	1000	620	20	13	15.58	8.64

Table 5.2. Base environmental (environment quality) indices of fossil fuel-fired power plants

Thermal power plant	Emissions t/GJ (fuel)			
	SO_2	NO_x	Particulates	CO_2
Fuel oil-fired, base	0.65	0.5	0.05	80
Fuel oil-fired, peak	0.65	0.5	0.05	80
Gas-fired, base	0	0.45	0.002	55
Gas-fired, peak	0	0.45	0.002	55
Coal-fired	0.6	0.4	0.85	95
Biomass-fired	0.01	0.2	0.045	95
Methanol-fired, base	0	0.45	0.002	65
Methanol-fired, peak	0	0.45	0.002	65
Hydrogen-fired, base	0	0.35	0	0
Hydrogen-fired, peak	0	0.35	0	0

Table 5.3. Technical and economic characteristics of nuclear power plants

Nuclear power plants	Year	Efficiency (net)	K, US$/kW	S, (US$/year) / kW	C_f, US$/GJ	C_{el}, Cent/kWh	C_{mod}, US$/GJ (Fuel)
Uranium-235- -fired	2025	0.33	2300	60	0.9	3.95	2.72
	2050	0.34	2300	60	0.9	3.92	2.80
	2075	0.35	2300	60	0.9	3.89	2.88
	2100	0.35	2300	60	0.9	3.89	2.88
Uranium-238- -fired	2025	0.35	3100	90	1	5.23	4.09
	2050	0.36	3100	90	1	5.20	4.20
	2075	0.38	3100	90	1	5.15	4.44
	2100	0.4	3100	90	1	5.10	4.67

Note: h = 7000 hour/year.

Table 5.4 presents HPP indices for the region of North America. This table presents efficiency, specific investments and operating costs based on the characteristics of transmission lines that connect HPP with electric power demand centers.

For other regions only the number of power utilisation hours h can be different.

The service life of the existing HPPs is 30 years, of the new ones — 50 years. Hydropower resources for these four types of HPP correspond to four cost categories indicated for ten regions in Table 4.35, Chapter 4.

Table 5.4. Technical and economic characteristics of HPP

Hydro power plants	Year	Efficiency (net)	h, h/year	K, US$/kW	S, (US$/year) / kW	$C_{el,}$ cent/kWh	$C_{mod,}$ US$/GJ (p.i.)
Existing	2025	0.85	4300	200	10	0.55	1.29
	2050	0.85	4300	300	10	0.70	1.65
	2075	0.85	4300	400	10	0.85	2.02
	2100	0.85	4300	500	10	1.01	2.38
New, cheap	2025	0.85	4200	2500	10	3.57	8.43
	2050	0.85	4200	2600	10	3.70	8.74
	2075	0.85	4200	2700	10	3.84	9.06
	2100	0.85	4200	2800	10	3.97	9.37
New, expensive	2025	0.84	4100	3500	15	5.14	12.00
	2050	0.84	4100	3600	15	5.28	12.32
	2075	0.84	4100	3700	15	5.42	12.64
	2100	0.84	4100	3700	15	5.42	12.64
New, very expensive	2025	0.83	3950	4500	20	6.88	15.87
	2050	0.83	3950	4600	20	7.03	16.20
	2075	0.83	3950	4700	20	7.17	16.52
	2100	0.83	3950	4800	20	7.71	16.85

5.2.4. Wind power plants (WPPs)

Wind energy has been used by people since ancient times. However its wider use for electric power generation began relatively recently, mainly in the last 20 years. By now many different types of wind power plants (WPPs) have been developed; the most efficient of them are plants with a horizontal axis of rotation consisting of the following main elements [108]: 1) a tower; 2) a head with a generator, a reducer and control system elements inside it; 3) a windwheel.

The main peculiarity that makes WPP different from traditional energy sources is variability of their capacity n, that changes depending on wind velocity v by the law

$$n(v) = Nf(v),$$

where N is an installed capacity of WPP (generator's capacity), f(v) is an operating characteristic (Figure 5.1). At a wind velocity of $v < V_0$ (usually $V_0 = 3 \div 5$ m/s) WPP stands idle; at $V_0 < v < V_1$ the capacity increases according to exponential dependence $f(v) \sim v^p$, where $p \approx 2 \div 3$; on the interval of velocities $V_1 < v < V_2$ the control system maintains capacity almost constant; and at high wind velocities ($v > V_2 = 20 \div 35$ m/s) WPP stops to avoid failures. Knowing the operating characteristic f(v) and probable distribution of wind velocity, it is possible to calculate the average capacity of WPP and the annual number of utilisation hours of the installed capacity N, and then to determine cost of the generated electric energy [121].

Since at a wind velocity $v < V_1$, WPP either stands idle or operates with a capacity lower than the rated one, for reliable power supply it has to be backed up by other energy sources, for example by those on fossil fuel. Wind-diesel systems that allow one to save expensive fossil fuel are applied to supply power (the consumed power is up to 10 MW) to small settlements located far from transmission lines and fuel supply sources. When estimating the economic efficiency of such a WPP, consideration must be given to the fact that some portion of power generated by it can be unused by a consumer (stored, etc.). Also the economic effect owing to partial replacement of the backing up capacity by WPP takes place when reliability of the power supply remains constant [122].

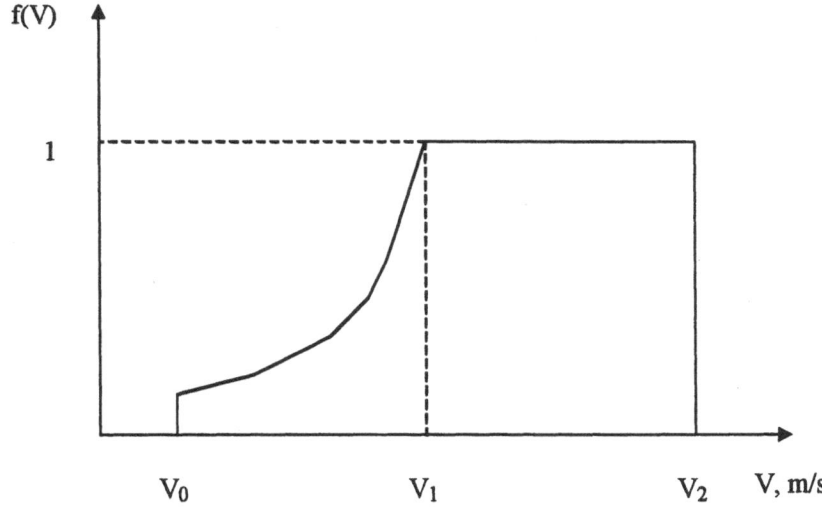

Figure 5.1. Operating characteristic of WPP.

The model GEM-10R does not consider independent power systems of small capacity since they consume a relatively small fraction of the total electric energy. The wind energy industry is supposed to develop on a large scale based on the wind power plants (WPP) that consist of installations with the capacity of $N=1 \div 1.5$ MW and are employed in large power systems.

Operation of wind power plants within a power system does not cause serious technical problems if their capacity is small compared to the capacity of the power system which, as a rule, is always observed: the maximum WPP fraction (about 3%) was reached in Denmark's energy sector. If the WPP fraction in power system is large (more than 20–30%) the network may suffer unacceptable frequency and voltage oscillations. However, progress in development of converters allows one to decrease WPP impact on electric networks and to eliminate this constraint.

Installed capacity of WPP N, kW, rated wind velocity V_1, m/s, and a windwheel diameter D, m, are connected by the relation

$$N=0.000481 \, \eta \, V_1^3 D^2,$$

where η is a wind energy utilisation factor equal to 0.4–0.5 (at a theoretically limiting value $\eta=0.593$) for the best WPP. According to this relation the windwheel diameter of a powerful WPP ($N \approx 1$ MW, $V_1 \approx 13$ m/s) is about 50 m.

The following classification of WPP operation mode is used depending on the blade design and methods of regulating the windwheel (turbine) rotation speed: 1) by changing the blade installation angle (α=var) and constant rotation frequency (ω=const); 2) by changing the blade installation angle (α=var) and alternative rotation frequency (ω=var); 3) with the constant (fixed) blade installation angle (α=const) and constant frequency (ω=const); 4) with the constant (fixed) blade installation angle (α=const) and alternative rotation frequency (ω=var). The WPP operation mode is optimal when at any value of wind velocity a rotation frequency ω changes in a way that the condition $\eta=\eta_{max}$ is met.

At present the total installed capacity of WPPs operating in the world makes up about 10 GW (Figure 5.2). Out of 2.1 GW capacity put into operation in 1998, 75% fell on European countries (Germany, Denmark, Great Britain, the Netherlands, Spain, Switzerland, Italy). For 5 years (1993–1998) the WPP capacity has increased almost four times; the annual capacity growth in Asia over the recent four years has amounted to 200–250 MW.

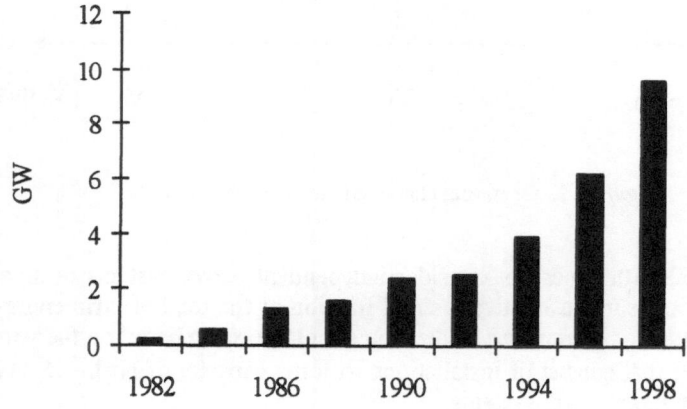

Figure 5.2. Installed capacity of WPP in the world.

The average capacity of installations produced by the industry grows steadily (Figure 5.3); at the same time the average specific energy output increases: from 150 kWh/m^2 of the windwheel area per year in 1980 to 500 kWh/m^2 in the late 1990s [123].

In 1982 the price of Danish wind power plants (without considering their construction) made up US$ 1770/kW; in 1997 it decreased up to US$ 850 kW. This decline was first of all due to a rise in unit capacity: the factor of cost decrease for installations of the same type amounted to 2–8% in 1982–1997 [123].

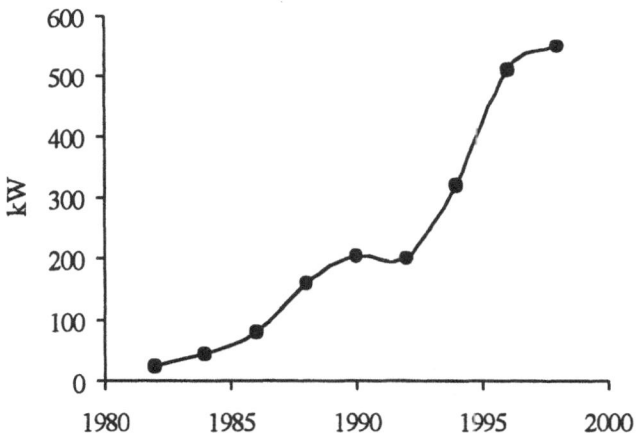

Figure 5.3. Average capacity of WPP.

At present the investments in WPP (with consideration for the costs of foundation and installation work) are in the range of US$ 1000–2500 kW and depend both on the rated capacity and on other parameters (windwheel diameter, height of the tower) [124]. In recent years installations with a capacity of 1.5 MW have appeared on the market. Their specific cost indices differ very little from the indices of installations with a capacity of 400–750 kW (capital investments of US$ 1000–1200/kW). This testifies to a relatively small possibility for the WPP efficiency increase compared to the achieved indices of the best modern installations.

Annual operating costs of WPP usually amount to 1.0 – 2.5 % of investments [125].

The cost of electric power generated by the modern WPP at an average long-term wind velocity V< 7 m/s reaches 3.5–4 cent/kWh. For prospective installations with optimised parameters it can decrease to 2.5–3 cent/kWh (Figure 5.4) [121, 123–125].

The indices characterising WPP operation, first of all the number of utilisation hours of the installed capacity, depend on wind conditions in the given area. Furthermore, transmission line construction should be envisaged since the areas with wind velocity suitable for WPP operation are often situated far from electric energy demand centers (in the model such costs are included directly in the WPP indices).

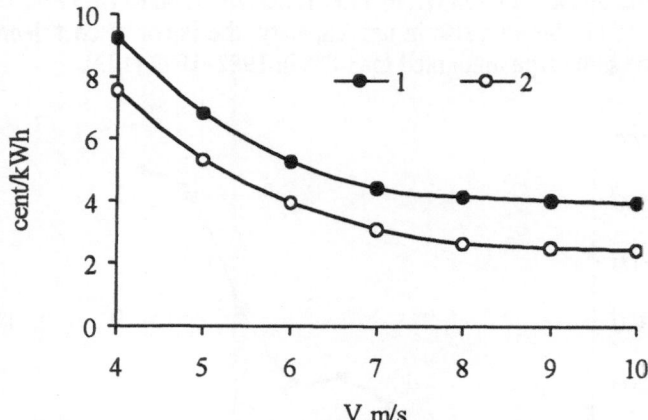

Figure 5.4. WPP electric energy cost. 1 — existing installations, 2 — perspective installations.

Table 5.5. Technical and economic characteristics of WPP

Wind power plants	Year	Efficiency (net)	h, h/year	K, US$/kW	S, (US$/year) / kW	C_{el}, cent/kWh	C_{mod}, US$/GJ (p.i.)
1	2025	0.20	2500	1100	25	4.21	2.34
	2050	0.25	2500	1000	23	3.84	2.66
	2075	0.30	2500	900	21	3.46	2.89
	2100	0.35	2500	900	21	3.46	3.37
2	2025	0.19	2500	1550	31	5.75	3.04
	2050	0.24	2500	1450	29	5.38	3.59
	2075	0.29	2500	1350	27	5.01	4.04
	2100	0.34	2500	1350	27	5.01	4.73
3	2025	0.20	1500	1100	25	7.01	3.89
	2050	0.25	1500	1000	23	6.39	4.44
	2075	0.30	1500	900	21	5.77	4.81
	2100	0.35	1500	900	21	5.77	5.61
4	2025	0.19	1500	1550	31	9.59	5.06
	2050	0.24	1500	1450	29	8.97	5.98
	2075	0.29	1500	1350	27	8.35	6.73
	2100	0.34	1500	1350	27	8.35	7.89

Thus, four technologies for converting wind energy into electrical energy were distinguished (each technology corresponds to specific wind energy resources — see Chapter 4, Table 4.39). They differ in wind velocity and remoteness from consumption centers (Table 5.5):

1) WPP–1 — the wind of classes 6–7 (an average wind velocity value for many years V=7 m/s), the distance L to a consumption center is less than 200 km (category 1);

2) WPP–2 — the same wind velocities, L=1000 km (categories 2 and 3);

3) WPP–3 — the wind of classes 3–4 (an average wind velocity value for many years V=5.4÷5.8 m/s), L< 200 km (categories 4 and 7);

4) WPP–4 — the same wind velocities, L=1000 km (categories 5, 6, 8 and 9).

The WPP's service life is 25 years.

5.2.5. Solar power plants (SPPs)

Electricity can be produced using solar energy either at thermal power plants that use the concentrated solar radiation flow or in the installations of direct energy conversion using photovoltaic converters (PVCs).

Thermal SPPs. By now many different types of thermal power plants have been suggested. The best known are thermal tower power plants with different solar energy concentrators that use either a steam turbine cycle (solar energy heats water or other working medium to the vaporous state, then steam goes to a turbine which rotates electric generator) or Stirling's engine (solar energy is used to heat a working medium in a special thermal engine that activates the rotor of the generator).

In the 1970s–1980s several pilot tower SPPs with a capacity of 0.5–10 MW were constructed in different countries. All of them were constructed on the same principle: mirrors-heliostats track the sun and reflect solar rays to a receiver (solar boiler unit producing steam that is forwarded to the turbine) installed on the top of the tower. Currently none of these SPPs is in operation due to low economic indices compared to conventional installations using fossil fuel.

Starting in the middle 1990s, nine solar power plants with parabolic trough concentrators with unit capacities of 14–80 MW (by LUZ technology) were constructed in Southern California. Tracking the sun along one axis, the concentrators focus solar radiation on tube receivers encased in vacuum tubes. There is a high temperature liquid heat carrier inside the receiver. It heats up and then gives the heat to the steam in the steam generator. Construction of such an SPP was stopped (turned out to be economically inefficient) after the federal budget of the USA had ceased to support their development. At present attempts to improve their efficiency are being made in the USA and Australia.

Solar power plants with a paraboloid concentrator tracking the sun along two axes is another SPP modification. Theoretically the paraboloid concentrator is the best. However, solar power plants with a paraboloid concentrator, unlike tower solar power plants and solar power plants with a parabolic trough concentrator, do not allow large unit capacities in one module due to the design peculiarities. Therefore they can be applied in isolated (independent) power systems.

Photovoltaic SPPs. Installations for direct conversion of solar radiation into electric power using photovoltaic converters (PVC) are of great interest. The main elements of PVC are crystals or film of semiconducting material where the energy of an absorbed light quantum is converted into electric power. The area of a single PVC is usually not large, therefore they are united into modules at SPP although additional power losses occur in connecting conductors.

The theoretical efficiency of a photovoltaic converter (photovoltaic cell) is determined by the width of a semiconductor's forbidden zone. The maximum efficiency is achieved at the width of the forbidden zone of 1.3–1.8 eV. Besides, to provide high PVC efficiency it is necessary to minimize external (optical) and internal (recombination of carriers, etc.) losses in it.

The maximum efficiency of SPP determined by the second law of thermodynamics is equal to 94.8%. The theoretical efficiency of the PVC of the first generation which is being intensively developed now, [homogeneous (crystalline and amorphous silicon) and heterogeneous (one heterojunction) photovoltaic cells] does not exceed 30%. The theoretical efficiency for the same types of PVC but with solar energy concentrators (PVC of the second generation) grows to 40%. The theoretical efficiency for PVC of the third generation (multiheterogeneous photovoltaic cells with solar energy concentrators, photovoltaic converters with dispersion of solar radiation in spectrum) is equal to 86.5% [126].

Before the 1980s the efficiency of mass-produced photovoltaic modules on amorphous silicon did not exceed 2–2.5%. In the early 1990s it made up 6–8% and by the middle 1990s increased up to 10–12% [127 – 129] (Figure 5.5).

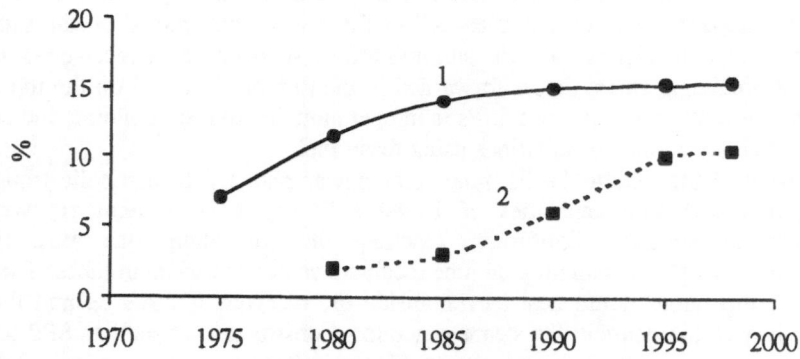

Figure 5.5. Dynamics of change in the solar modules' efficiency. 1— on crystalline silicon, 2— on amorphous silicon.

At present the efficiency of more expensive mass-produced PVC on crystalline silicon is 16% [152]. The pilot samples provide the efficiency of 20–21% [126].

Hopes for improvement of photovoltaic cell efficiency should apparently be associated with the cascade heterogeneous photovoltaic converters (having 2–3 or even 4 heterojunctions). At present photovoltaic cells with efficiency of 29–30% (cascade heterophotovoltaic converters based on gallium arsenide with concentrators) have been generated in laboratory conditions [130]. At present such cells are too expensive but their cost is supposed to considerably decrease by the middle of the 21st century. The efficiency of such PVC that can really be achieved is 35–40% [126].

In the middle 1980s about 150 MW of PVC were in operation in the world. The overwhelming majority of them operated on crystalline silicon. The total sales of solar cells made up 100 MW in 1997 [128], 150 MW — in 1998, including 70 MW on polycrystalline silicon, 60 MW — on crystalline silicon, 19 MW — on amorphous silicon [127].

The most important advantages of photovoltaic SPPs are their module design, high degree of factory readiness, the simplicity of service, high reliability, no harmful emissions into the environment during operation and no special requirements to the site (SPP can be located at the most inconvenient places). Strong dependence of solar power plants on weather conditions, large areas of the land required, direct current generation, and high cost should be considered the main disadvantages.

The specific cost of photovoltaic cells amounted to more than US$ 30 000/kW in 1974, about US$ 15000/kW — in 1984, by the mid-1990s it decreased to US$ 3500/kW [128, 129] (Figure 5.6).

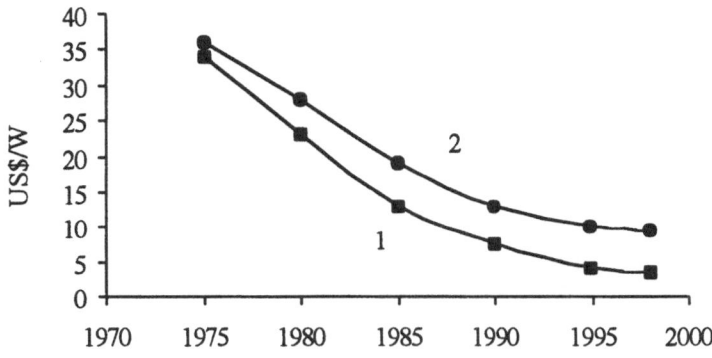

Figure 5.6. Price of solar modules. 1— large modules, 2 — small modules.

The photovoltaic SPP in a general case includes photovoltaic modules installed in fixed or rotating arrays; concentrators of solar radiation (for PVC with concentrators); DC-to-DC inverters; DC-to-AC inverters; step-up transformers; power storage systems.

The voltage-amperage characteristic and the point of maximum capacity of the photovoltaic converter are progressively changing due to changes in solar radiation capacity and PVC temperature. To provide PVC operation at a maximum capacity point the controlled DC-to-DC inverter is used. Direct current generated by PVC is converted into alternating current by an inverter with subsequent filtration of higher harmonics.

The solar module cost usually makes up 50–60% or more of the total cost of the pilot SPPs in operation.

Table 5.6 presents the forecasted estimates of SPP indices accepted for the calculations on GEM-10R. They suppose the use of photovoltaic SPPs whose updating is expected as a result of technological progress. Thus, following the logic of presenting solar energy resources in terms of geographical and climatic characteristics of an area (see Chapter 4) three classes of SPPs were distinguished: 1) cheap SPP located in regions with a high level of solar radiation, 2) expensive SPP located in regions with an average level of solar radiation, 3) very expensive SPP located in regions with a low level of solar radiation. The SPP electric energy cost grows due to decrease of utilisation hours of the installed capacity h. Table 5.6 presents SPP characteristics based on the costs of electric energy transmission to the consumption centers and corresponding losses of transmission lines. The cheap SPP located in regions with high insolation (usually these regions are unpopulated) are the most remote from large consumption centers. Service life of SPP is accepted equal to 30 years; the energy consumption for auxiliaries is 4%.

Table 5.6. Technical and economic characteristics of SPP

Solar power plants	Year	Efficiency (net)	h, h/year	K, US$/kW	S, (US$/year) / kW	C_{el}, cent/kWh	C_{mod}, US$/GJ (p.i.)
Cheap	2025	0.14	2200	2450	16	8.30	3.23
	2050	0.24	2200	1950	16	6.76	4.51
	2075	0.29	2200	1750	16	6.15	4.95
	2100	0.34	2200	1650	16	5.84	5.52
Expensive	2025	0.15	1650	2250	13	10.06	4.19
	2050	0.25	1650	1750	13	8.01	5.56
	2075	0.30	1650	1550	13	7.19	5.99
	2100	0.35	1650	1450	13	6.78	6.59
Very expensive	2025	0.15	1100	2000	10	13.27	5.53
	2050	0.25	1100	1500	10	10.19	7.07
	2075	0.30	1100	1300	10	8.96	7.46
	2100	0.35	1100	1200	10	8.34	8.11

Figure 5.7 presents the cost estimates of electric energy generated by SPP at different levels of solar radiation. As illustrated in this figure and Table 5.6. the cost of electric energy generated by SPPs under the best climatic conditions will amount to about 8 cent/kWh in 2025 and will decrease up to about 6 cent/kWh in 2100.

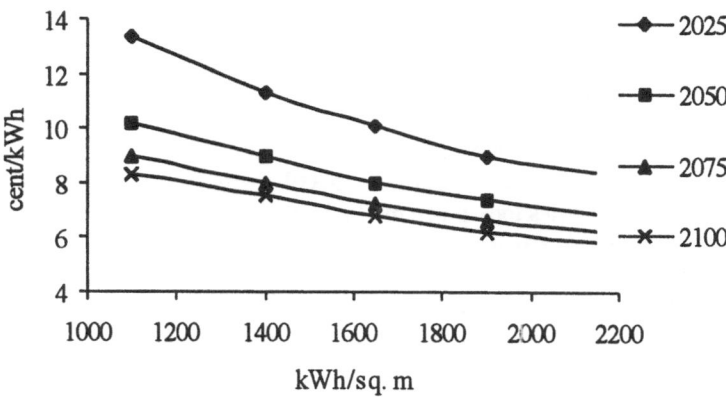

Figure 5.7. Cost of electric energy of solar power plants for different levels of solar radiation.

5.2.6. Geothermal power plants (GeoTPP)

Italy is considered to be a country of native geothermal energy. In 1904 in the province of Toskana, electric energy was generated using natural steam for the first time. The first industrial geothermal power plant in the world, Larderello–1 with a capacity of 12 MW, constructed in 1916, is still in operation.

The total capacity of GeoTPP in the world amounted to about 6.5 GW in the middle 1990s against 5.3 GW in 1988 [118] and 0.5 – 0.8 GW in the 1960s (Figure 5.8) [131]. If all the plants already planned are commissioned, that capacity could exceed 10 GW in the beginning of the 21st century.

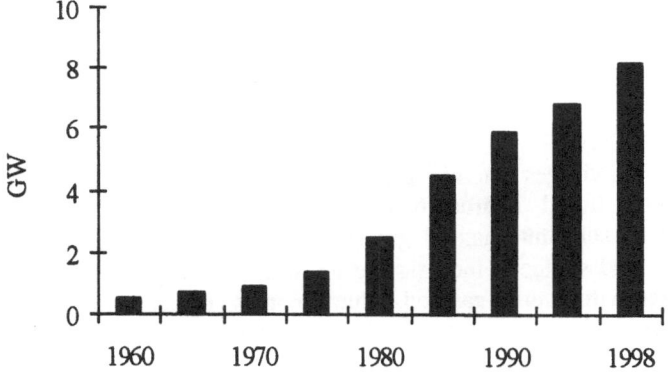

Figure 5.8. Total installed capacity of GeoTPP.

According to data of the Internet site of the International Geothermal Association, in 1988 the installed capacities of GeoTPPs were distributed as follows: 37% were in Asia (including 1.85 MW in the Philippines, 0.53 MW in Japan, 0.59 GW in Indonesia), 35% were in North America (2.85 MW in the USA), more than 11% — in Europe (including 0.77 GW in Italy, 0.14 GW in Iceland), about 13% in Latin America (including 0.74 MW in Mexico, 0.12 MW in Costa Rica, 0.1 GW in El Salvador), 4% — in others (including 0.35 GW in New Zealand). The largest growth of GeoTPP capacities took place in 1980–1984 (13.5% a year) [131]. In later years these growth rates dropped.

Table 5.7 presents technical and economic indices of GeoTPP used for calculations by model GEM-10R. In this application of the model the GeoTPPs were supposed to have no significant potential for improvement due to the low parameters of geothermal energy used.

Table 5.7. Technical and economic characteristics of GeoTPP in 2025–2100

GeoTPP	K, US$/kW	S, (US$/year)/kW	C_{el}, cent/kWh	C_{mod}, US$/GJ (p.i.)
Cheap	3300	120	7.78	5.41
Expensive	4500	140	10.21	7.09
Very expensive	5500	180	12.96	9.00

Note: h=5000 hour/year, efficiency (net) = 0.25.

The GeoTPP classes mentioned in Table 5.7 correspond to classification of the economic factors of geothermal resources (see Table 4.43). Cheap GeoTPP uses high temperature thermal waters, the expensive one uses geothermal anomalies at great depths, a very expensive one uses the heat of dry rocks (at great depths as well).

5.3. Technologies for production of secondary chemical energy carriers

GEM-10R envisages potential production of the following secondary chemical energy carriers: liquid hydrocarbons of light and heavy fractions, gaseous hydrocarbons — substitute natural gas (SNG), methanol, hydrogen and synthetic biomass. SNG and synthetic biomass are not used as independent energy carriers. They are mixed with natural gas and natural biomass respectively. This is possible because they have the properties of their natural analogues. This is also expedient since it makes the model's dimensionality smaller.

These carriers differ greatly in transport properties. Thus their costs must be added to the main costs for the corresponding technology when calculating costs of the transport infrastructure.

Tables 5.8–5.11 present the technical and economic characteristics of technologies for production of secondary chemical energy carriers assumed in GEM-10R. The notations used are: efficiency — a fraction of the total output of refining products; h — the number of utilisation hours of the installed capacity of the refinery, h/yr; K_{ref} — specific investments in refining, US$/kWh of the secondary energy (s.e.); S_{ref} — specific operating costs at refining, US$/(kWh); K_{tr} — specific investments in transport infrastructure, US$/kW (s.e.); S_{tr} — specific operating costs of transport infrastructure operation, US$/(kW/yr); C_f — fuel cost (conditional - to determine $C_{s.e.}$), US$/GJ; $C_{s.e.}$ — cost of the secondary energy carrier production, US$/GJ (s.e.); C_{mod} — specific (reduced) costs of energy carrier production (without fuel constituent, in terms of the power input) presented in the model's functional, US$/GJ (p.i.).

5.3.1. Liquid and gaseous hydrocarbon production

Four technologies of liquid hydrocarbon production are envisaged in the GEM-10R model:

1) a conventional oil refinery producing motor fuels (MF) and fuel oil in proportion of 0.7:0.15 (oil product output amounts to 0.85);

2) oil refinery with a deep fuel oil refining that produces only motor fuels;

3) coal conversion into liquid hydrocarbons of light fractions;

4) methanol conversion into liquid hydrocarbons of light fractions .

All the technologies except for the third one were supposed to reach maximum updating by 2025 and then their technical and economic indices will remain constant. The third technology — coal conversion into liquid hydrocarbon of light fractions — will be constantly improved. At present the development goes in two directions: coal hydrogenation and synthesis of liquid hydrocarbons from CO-hydrogen mixture. Now it is impossible to give preference to either of them — both have advantages and disadvantages. Therefore Table 5.8 presents the indices common ("synthetic") to them [62].

Two macrotechnologies of SNG production are envisaged as well: from coal and biomass. Currently three principally different technological processes of SNG production from solid fuels are being developed: 1) hydrogasification, 2) synthesis from CO-hydrogen mixture, 3) biological decomposition under anaerobic conditions. These technologies have great potential for improvement [62]. Table 5.8 summarizes their indices as applied to coal and biomass usage. The service life of each of the technologies was taken equal to 30 years.

Table 5.8. Technical and economic characteristics of technologies for liquid and gaseous hydrocarbon production

Conversion technology	Year	Effici-ency	K_{ref}, US$/ kW	S_{ref}, (US$/yr) kW	K_{tr}, US$/ kW	S_{tr}, (US$/yr) kW	C_f, US$/ GJ	$C_{s.e.}$, US$/ GJ (s.e.)	C_{mod}, US$/GJ (p.i.)
Oil into	2025	0.85	600	30	8	0.5	2.0	5.35	2.54
MF+fuel oil	2050	0.85	600	30	8	0.5	3.0	6.52	2.54
(MF:fuel	2075	0.85	600	30	8	0.5	4.0	7.70	2.54
oil	2100	0.85	600	30	8	0.5	5.0	8.88	2.54
=0.7:0.15)									
Oil into	2025	0.77	780	39	8	0.5	2.0	6.48	2.99
MF	2050	0.77	780	39	8	0.5	3.0	7.77	2.99
	2075	0.77	780	39	8	0.5	4.0	9.07	2.99
	2100	0.77	780	39	8	0.5	5.0	10.37	2.99
Coal into	2025	0.55	1520	70	8	0.5	1.5	9.99	3.99
MF	2050	0.57	1520	70	8	0.5	2.0	10.77	4.14
	2075	0.59	1520	70	8	0.5	2.4	11.33	4.28
	2100	0.61	1520	70	8	0.5	2.7	11.69	4.43
Coal into	2025	0.65	1050	55	45	4.0	1.5	7.87	3.62
SNG	2050	0.68	1050	55	45	4.0	2.0	8.51	3.78
	2075	0.71	1050	55	45	4.0	2.4	8.95	3.95
	2100	0.73	1050	55	45	4.0	2.7	9.26	4.06
Biomass	2025	0.66	950	53	45	4.0	1.9	8.08	3.43
into SNG	2050	0.70	950	53	45	4.0	2.3	8.49	3.64
	2075	0.73	950	53	45	4.0	2.6	8.76	3.80
	2100	0.75	950	53	45	4.0	3.5	9.87	4.90

Note. $h = 6500$ h/yr.

5.3.2. Methanol production

Methanol is one of the most universal energy carriers. It can be efficiently applied to produce all the final energy forms without exception. Methanol is a base energy carrier in a concept of a "horizontally integrated energy system" [58, 62, 132].

The model GEM-10R presents five methanol production technologies: one — from oil, natural gas and biomass and two — from coal (Table 5.9). The processes of feedstock gasification, synthesis gas cleaning and subsequent synthesis of methanol are the basis for all the technologies. All of them have the potential to develop. The service life of these technologies was taken to be 30 years as well.

5.3.3. Hydrogen production

Hydrogen is an environmentally friendly, universal energy carrier. It can be applied to generation of electrical, thermal and mechanical energy. It is precisely hydrogen in which we place our hopes for introduction of fuel elements into the electric power industry and the transport. Hydrogen is the main energy carrier in a concept of "the hydrogen energy system"[62, 133].

Table 5.9. Technical and economic characteristics of methanol production technologies

Conversion technology	Year	Efficiency	K_{ref}, US$/ kW	S_{ref}, (US$/yr) kW	K_{tr}, US$/ kW	S_{tr}, (US$/yr) kW	C_f, US$/ GJ	$C_{s.e.}$, US$/ GJ (s.e.)	C_{mod}, US$/GJ (p.i.)
Oil into	2025	0.65	830	43	8	0.5	2.0	7.27	2.72
methanol	2050	0.67	830	44	8	0.5	3.0	8.71	2.83
	2075	0.69	830	45	8	0.5	4.0	10.07	2.95
	2100	0.70	830	45	8	0.5	5.0	11.42	2.99
Gas into	2025	0.65	750	38	45	4.0	2.3	7.54	2.60
methanol	2050	0.67	750	39	45	4.0	3.3	8.97	2.71
	2075	0.69	750	40	45	4.0	4.3	10.32	2.82
	2100	0.70	750	40	45	4.0	5.3	11.66	2.86
Coal into	2025	0.60	1440	64	8	0.5	1.5	9.28	4.11
methanol	2050	0.62	1450	65	8	0.5	2.0	10.08	4.25
	2075	0.64	1460	66	8	0.5	2.4	10.67	4.39
	2100	0.65	1470	67	8	0.5	2.7	11.15	4.45
Coal +	2025	1.30	1440	64	8	0.5	1.5	9.11	5.45
hydrogen	2050	1.35	1450	65	8	0.5	1.5	8.97	4.25
into	2075	1.38	1460	66	8	0.5	1.5	8.89	4.39
methanol	2100	1.40	1470	67	8	0.5	1.5	8.84	4.45
(coal: hydrogen =1:0.8)									
Biomass	2025	0.60	1250	60	8	0.5	1.9	9.25	3.65
into	2050	0.62	1250	60	8	0.5	2.3	9.79	3.77
methanol	2075	0.64	1250	60	8	0.5	2.6	10.15	3.89
	2100	0.65	1250	60	8	0.5	3.5	11.47	3.95

Note: h=6500 hr/yr.

The model GEM-10R suggests ten hydrogen production technologies (Table 5.10). Hydrogen is produced using fossil fuels by the scheme "feedstock gasification — producer gas cleaning — reaction of shift (CO conversion into hydrogen)".

Using renewable energy resources (solar energy and energy from space) hydrogen is produced during the electrolysis process. Hence, in this case electric energy has already been generated. However, in this process there is no need to convert direct current to alternating current and vice versa. Thus, some costs are saved owing to the combined hydrogen production technology based on renewables as compared to separate electric energy production and subsequent electrolysis. In the future the acceptable indices will likely be achieved in the technologies of direct photochemical water decomposition.

There are two main methods of nuclear energy based hydrogen production: electrolysis and direct high-temperature thermochemical water decomposition. The first method is technically developed, which can not yet be said about the second method. At the same time the second method has a significant potential for improvement. Table 5.10 presents the indices for the first method. The service life of each of the technologies was taken equal to 30 years.

Table 5.10. Technical and economic characteristics of hydrogen production technologies

Conversion technology	Year	Efficiency	h, h/yr	K_{ref}, US$/ kW	S_{ref}, (US$/ yr)/ kW	K_{tr}, US$/ kW	S_{tr}, (US$/yr) /kW	C_f US$/ GJ	$C_{s.e.}$, US$/ G (s.e.)	C_{mod} US$/G (p.i.)
Oil into	2025	0.65	6500	860	45	68	6	2.0	7.84	3.09
hydrogen	2050	0.68	6500	860	46	68	6	3.0	9.21	3.27
	2075	0.70	6500	860	47	68	6	4.0	10.56	3.39
	2100	0.72	6500	860	47	68	6	5.0	11.79	3.49
Gas into	2025	0.65	6500	780	40	68	6	2.3	7.86	2.81
hydrogen	2050	0.68	6500	780	41	68	6	3.3	9.22	2.97
	2075	0.70	6500	780	42	68	6	4.3	10.55	3.09
	2100	0.72	6500	780	42	68	6	5.3	11.77	3.17
Coal into	2025	0.58	6500	1500	70	68	6	1.5	10.19	4.03
hydrogen	2050	0.60	6500	1500	70	68	6	2.0	10.94	4.34
	2075	0.62	6500	1500	70	68	6	2.4	11.48	4.56
	2100	0.63	6500	1500	70	68	6	2.7	11.89	4.72
Uran-235	2025	0.30	6500	4150	100	68	6	0.9	19.26	4.88
into	2050	0.35	6500	4150	100	68	6	0.9	18.83	5.69
hydrogen	2075	0.38	6500	4150	100	68	6	0.9	18.62	6.18
	2100	0.40	6500	4150	100	68	6	0.9	18.51	6.50
Uran-238	2025	0.29	6500	5500	110	68	6	1.0	23.88	5.93
into	2050	0.31	6500	5500	110	68	6	1.0	23.66	6.34
hydrogen	2075	0.34	6500	5500	110	68	6	1.0	23.38	6.95
	2100	0.39	6500	5500	110	68	6	1.0	23.00	7.97
Biomass	2025	0.58	6500	1310	66	68	6	1.9	10.18	4.01
into	2050	0.60	6500	1310	66	68	6	2.3	10.74	4.14
hydrogen	2075	0.62	6500	1310	66	68	6	2.6	11.10	4.28
	2100	0.63	6500	1310	66	68	6	3.5	12.46	4.35
Solar	2025	0.11	2200	2450	30	100	9	0.0	25.87	2.85
energy	2050	0.18	2200	1970	30	100	9	0.0	21.93	3.95
(cheap)	2075	0.23	2200	1790	30	100	9	0.0	20.45	4.70
into hydrogen	2100	0.29	2200	1700	30	100	9	0.0	19.71	5.72
Energy	2025	0.53	2800	5800	230	68	6	0.0	61.28	32.5
from Space	2050	0.56	2800	5300	210	68	6	0.0	56.07	31.4
into	2075	0.60	2800	4800	190	68	6	0.0	50.86	30.5
hydrogen	2100	0.63	2800	4300	170	68	6	0.0	45.65	28.8
Electric	2025	0.71	4000	450	19	68	6	11.0	19.57	2.89
energy	2050	0.73	4000	450	19	68	6	12.0	20.51	2.98
into	2075	0.75	4000	450	19	68	6	12.5	20.74	3.06
hydrogen	2100	0.77	4000	450	19	68	6	13.0	20.96	3.14

5.3.4. Synthetic biomass production in water reservoirs

The GEM-10R model offers three technologies of synthetic biomass production by using solar energy (Table 5.11).

Table 5.11. Technical and economic indices of synthetic biomass production technologies

Technology for converting solar energy into biomass	Year	Effici-ency	h, h/yr	K_{ref}, US$/ kW	S_{ref}, (US$ /yr)/ kW	K_{tr}, US$/ kW	S_{tr}, (US$ /yr)/ kW	$C_{s.e.}$, US$/ GJ (s.e.)	C_{mod}, US$/GJ (p.i.)
Cheap	2025	0.05	2200	1200	80	70	20	23.06	1.15
	2050	0.07	2200	1200	80	70	20	23.06	1.61
	2075	0.10	2200	1200	80	70	20	23.06	2.31
	2100	0.12	2200	1200	80	70	20	23.06	2.77
Expensive	2025	0.03	1650	900	75	60	17	26.00	0.78
	2050	0.035	1650	900	75	60	17	26.00	0.91
	2075	0.04	1650	900	75	60	17	26.00	1.04
	2100	0.04	1650	900	75	60	17	26.00	1.04
Very expensive	2025	0.02	1100	800	70	50	15	35.43	0.71
	2050	0.025	1100	800	70	50	15	35.43	0.89
	2075	0.03	1100	800	70	50	15	35.43	1.06
	2100	0.03	1100	800	70	50	15	35.43	1.06

Note: $C_f = 0$.

Carbon dioxide from the atmosphere serves as a "raw material". The process of photosynthesis is performed in a specially made reservoir that provides the highest biomass increase. A repeated efficiency enhancement is supposed by the end of the century. Biomass cost depends on the annual number of radiation hours h; the cost increases as the number of hours utilised decreases.

5.4. Technologies for production of final energy forms

5.4.1. Thermal energy generation

Table 5.12 presents forecasted technical and economic indices of technologies for thermal energy generation that are included in the GEM-10R model. Emphasis should be made on the most important distinguishing feature of thermal energy production technologies — they are abundant. They differ, firstly, in their purpose (thermal energy generation of high, medium and low potentials) and secondly in a unit capacity (from parts to hundreds of megawatts). These factors can be taken into account in the model without any principal problems except for a considerable increase in dimensionality.

The experience of applying system models [62] showed that it is quite appropriate to use highly aggregated technologies ("macrotechnologies") in long-term energy studies. Table 5.12 presents generalised characteristics of thermal energy generation macrotechnologies for each fuel kind, averaged for a region and operating conditions for a year.

Table 5.12. Technical and economic characteristics of thermal energy generation technologies

Fuel (energy)	Year	Effici-ency	K_i, US$/ kW(t.e.)	K_{tr}, US$/ kW(p.i.)	S_i, (US$/yr) kW(t.e.)	S_{tr}, (US$/yr) kW(p.i.)	C_f, US$/ GJ	$C_{s.e.}$, US$/ GJ (t.e.)	C_{mod}, US$/GJ (p.i.)
Gas	2025	0.75	400	35	6	3.5	2.3	5.83	2.07
	2050	0.80	400	35	6	3.5	3.3	6.85	2.18
	2075	0.83	400	35	6	3.5	4.3	7.89	2.25
	2100	0.85	400	35	6	3.5	5.3	8.93	2.29
Fuel oil	2025	0.75	400	20	6	3.0	2.0	5.29	1.97
	2050	0.80	400	20	6	3.0	3.0	6.35	2.08
	2075	0.83	400	20	6	3.0	4.0	7.40	2.14
	2100	0.85	400	20	6	3.0	5.0	8.46	2.19
Coal	2025	0.70	650	25	11	4.0	1.5	6.40	2.98
	2050	0.75	650	25	11	4.0	2.0	6.89	3.17
	2075	0.78	650	25	11	4.0	2.4	7.28	3.28
	2100	0.80	650	25	11	4.0	2.7	7.56	3.35
Biomass	2025	0.70	660	28	12	4.0	1.9	7.11	3.07
	2050	0.75	660	28	12	4.0	2.3	7.42	3.27
	2075	0.73	660	28	12	4.0	2.6	7.93	3.19
	2100	0.80	660	28	12	4.0	3.5	8.70	3.46
Methanol	2025	0.75	400	20	6	3.0	9.0	14.62	1.97
	2050	0.80	400	20	6	3.0	10.0	15.10	2.08
	2075	0.83	400	20	6	3.0	11.0	15.84	2.14
	2100	0.85	400	20	6	3.0	12.0	16.69	2.19
Hydrogen	2025	0.75	400	50	6	5.0	10.0	16.32	2.24
	2050	0.80	400	50	6	5.0	11.0	16.69	2.35
	2075	0.83	400	50	6	5.0	12.0	17.37	2.42
	2100	0.85	400	50	6	5.0	13.0	18.19	2.46
Uranium-235	2025	0.80	1500	0	55	0.0	0.9	11.72	8.48
	2050	0.80	1500	0	55	0.0	0.9	11.72	8.48
	2075	0.80	1500	0	55	0.0	0.9	11.72	8.48
	2100	0.80	1500	0	55	0.0	0.9	11.72	8.48
Electric energy	2025	0.90	250	90	3	2.5	10.0	13.09	1.78
	2050	0.90	250	90	3	2.5	11.0	14.20	1.78
	2075	0.90	250	90	3	2.5	12.0	15.32	1.78
	2100	0.90	250	90	3	2.5	13.0	16.43	1.78

Note. h=4000 h/year.

The notations used in Table 12 are: efficiency — thermal energy generation efficiency with regard to losses of power (fuel) input at the distribution stage; h — number of utilisation hours of the installed capacity, h/yr; K_i — specific investments in technological installation (boiler plant), US$/kW (thermal energy — t.e.); S_i — specific operating costs of thermal energy production, US$/(kW/yr); K_{tr} — specific investments in transport infrastructure, US$/kW (t.e.); S_{tr} — specific operating costs of transport infrastructure, US$ (kW/yr); C_f — fuel cost (conditional), US$/GJ (t.e.); C_{te} — thermal energy generation cost, US$/GJ (t.e.); C_{mod} — thermal energy generation cost in the model (without fuel constituent, in terms of power input), US$/GJ (p.i.).

Forecasting the indices of thermal energy production technologies based on fossil fuels and hydrogen envisages potentialities for considerable enhancement of their energy efficiency. It is connected, firstly, with application of more efficient heat recuperation systems, secondly, with deeper cooling of exhaust gases (including water steam condensation which is technologically acceptable, first of all, for "clean" fuels — hydrogen, methanol and natural gas) and, thirdly, with wider application of automatic systems to control technological processes in heat supply systems including combustion processes.

5.4.2. Mechanical energy production

The GEM-10R model presents five macrotechnologies of mechanical energy production. Table 5.13 presents their generalised technical and economic indices.

Table 5.13. Technical and economic indices of mechanical energy production technologies

Fuel	Year	Effici-ency	K_m, US$/kW (m.e.)	K_{tr}, US$/ kW(p.i.)	S_m, (US$/yr) kW(m.e.)	S_{tr}, (US$/yr) kW(p.i.)	C_f, US$/GJ	$C_{m.e.}$, US$/ GJ (m.e.)	C_{mod}, US$/GJ (p.i.)
Motor	2025	0.25	200	20	5.6	4.0	7.0	35.1	1.76
fuel	2050	0.26	200	20	5.6	4.0	8.0	37.8	1.83
	2075	0.28	200	20	5.6	4.0	9.0	39.2	1.98
	2100	0.30	200	20	5.6	4.0	10.0	40.4	2.12
Natural	2025	0.23	270	40	6.7	7.0	2.3	20.0	2.29
gas	2050	0.24	270	40	6.7	7.0	3.3	23.7	2.39
	2075	0.26	270	40	6.7	7.0	4.3	26.5	2.59
	2100	0.28	270	40	6.7	7.0	5.3	28.9	2.79
Methanol	2025	0.25	230	25	6.0	4.5	9.0	44.1	2.01
	2050	0.26	230	25	6.0	4.5	10.0	46.5	2.10
	2075	0.28	230	25	6.0	4.5	11.0	47.3	2.26
	2100	0.30	230	25	6.0	4.5	12.0	48.1	2.42
Hydrogen	2025	0.22	320	70	9.5	11.0	10.0	58.6	2.89
	2050	0.23	320	70	9.5	11.0	11.0	61.0	3.02
	2075	0.25	320	70	9.5	11.0	12.0	61.1	3.29
	2100	0.27	320	70	9.5	11.0	13.0	61.3	3.55
Electric	2025	0.50	370	80	4.4	6.0	11.0	34.7	6.36
energy	2050	0.54	370	80	4.4	6.0	12.0	34.9	6.87
	2075	0.57	370	80	4.4	6.0	12.5	34.6	7.25
	2100	0.60	370	80	4.4	6.0	13.0	34.4	7.63

Note. h = 1500 h/year.

Since various energy carriers used for mechanical energy production differ greatly in costs of construction and operation of the corresponding transport infrastructure, these costs are taken into account in the model and reflected in Table 5.13. The economic characteristics of mechanical energy production technology are: K_m — specific investments in technological installation, US$/kW (mechanical energy — m.e.); S_M — specific operating costs at mechanical energy production, US$/ (kW/yr); $C_{m.e.}$ — cost of mechanical energy production, US$/GJ (m.e.). The remaining notations are the same as in Table 5.12.

The technical and economic indices of mechanical energy production technologies may be improved, first of all, by updating the engine designs, secondly, by bettering the combustion process characteristics and, thirdly, by applying new technologies, in particular, fuel element technologies.

5.4.3. Chemical energy production

The base raw material for chemical energy production assumed in the GEM-10R model is liquid hydrocarbons of light fractions (the products of oil refining and coal conversion). They can be replaced by gas (natural or SNG) and methanol. This is associated with certain costs. The additional costs of raw material substitution when producing chemical energy are given in Table 5.14. Here K_c — additional specific investments in reconstruction of technological installations for the new raw material, US$/kW (chemical energy — c.e.); S_c — additional specific operating costs at chemical energy production in the case of conversion to a new raw material, US$/(kW/yr).

All the indices of chemical energy production technologies are taken constant in time.

Table 5.14. Technical and economic characteristics of chemical energy production technologies

Raw material	Year	Efficiency	K_c, US$/kW (c.e.)	K_{tr}, US$/kW (p.i.)	S_c, (US$/yr) /kW(c.e.)	S_{tr}, (US$/yr) /kW(p.i..)	C_{mod}, US$/GJ (p.i.)
Liquid hydrocarbons	2025–2100	1	0	0	0	0	0.000
Gas	2025–2100	0.85	80	4	10	0.5	0.739
Methanol	2025–2100	0.93	50	2.5	0	0	0.158

Note. h = 6000 h/yr.

5.5. Space power systems

The major projects for creation of Space power systems that use solar energy and transmit it to the Earth using super high-frequency (SHF) or microwave radiation involve Solar power satellites (SPS) in geostationary orbit (spinning with rotation velocity of the Earth and "hanging" above the receiving antenna — rectenna) with a capacity of about 5 GW [135] and Lunar power system (LPS) with a capacity of 20000 GW [134]. Another large-scale project supposing helium–3 production on the Moon with its transportation to the Earth and further utilisation in thermonuclear reactors [136] can not be considered a Space power system (thermonuclear power plants are located on the Earth). The other known proposals are of smaller scale and can be taken as intermediate stages in implementation of the first two projects.

A comprehensive discussion of the related problems [137] showed that these projects could be technically implemented when producing the main elements of SPS and LPS from lunar materials on the Moon. For this purpose it is necessary to create manned lunar bases and maximum computerised and robotised production systems, including fuel production for rockets. The calculations showed that production of SPS and LPS elements on the Earth with their further launching into space is absolutely impossible from the economic and environmental viewpoints. A main argument in favor of appropriate studies and experiments consists in the fact that humanity will undoubtedly develop space which makes the statement of goals associated with energy supply to the Earth quite reasonable.

The initial conception of a Lunar power system [135] has some disadvantages (they will be pointed out below), therefore the authors of the project suggested other kinds of LPS [138] for consideration. The emphasis will be made on (using [9–11]):

— potential constraints on unit capacities of rectennas and their location on the territory of the Earth;

— possible modes of power supply from space to the Earth (interrupted or uninterrupted);

— economic indices, determining efficiency and competitiveness of power from space.

Solar power satellites (SPS) (Figure 5.9). In fact there is only one geostationary (or geosynchronous) orbit (GSO), that passes exactly above the equator at a height of about 35 thousand km from the Earth's surface and provides SPS "hanging up" above an equator point. It is used for many other satellites, particularly for teleradio communication. This will undoubtedly lead to a constrained total number of SPSs that could be located in GSO. Besides, it will be possible to receive energy from them only in a certain "equatorial" zone (in polar zones of the Earth the satellites in GSO will not be seen).

It is quite obvious that there will also be a constraint on the potential capacity of each power satellite. Now, the SPS capacity is assumed equal to 5 GW (see, for example, [139]). In this case the size of a solar collector will make up about 50 km2. In the process of SPS spinning around the Earth this collector will have to be permanently oriented at the Sun and the microwave antenna (with a diameter of about 1 km), fixed at either of the collector's ends, — at one and the same point (rectenna) on the Earth. The possibility of creating a satellite of such a size with orientation mechanisms will apparently be determined only in the process of creating a smaller pilot SPS.

Thus, when supplying power from Solar power satellites to the Earth there will be constraints on unit capacities of SPS, their total number (and, hence, on the total scale of this Space system) and on location of receiving antennas. Concrete quantitative values (sizes) of these constraints will be determined in the process of further studies and tests. Tentatively it may be supposed that about a hundred of 5 GW SPSs (totally 500 GW) can be constructed with rectennas in the zone up to 60^{0} north/south latitude.

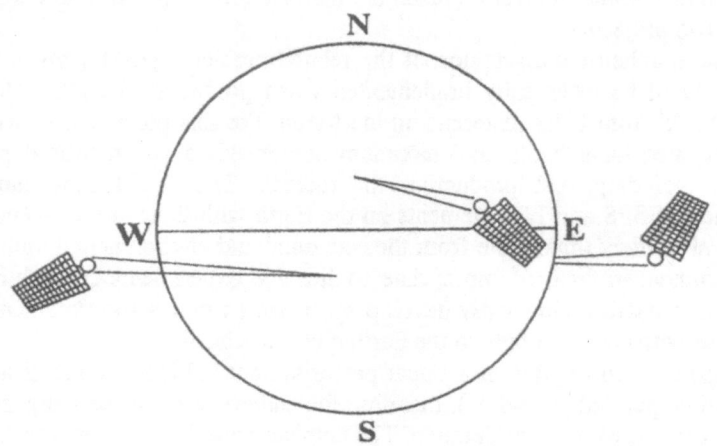

Figure 5.9. Block diagram of SPS.

Another constraint on SPS application may arise in terms of polluting the Earth-orbital space with SPS elements that have served their lifetime. Given the size and potential lifetime (20–40 years) of each satellite, it is clear that it will be hard to remove an obsolete SPS from orbit if it has been created on a large scale (hundreds of GW) with a long-term (centuries) power supply.

As to the modes of receiving energy from SPS, they seem to be quite favorable. Most of the year the satellites in GSO will be exposed to Sun's light and can continuously supply rectennas with power. Only twice a year, at the period of the equinox (when the Sun moves along the equator) will SPS be in the shadow of the Earth for about one hour at midnight for several weeks. This will lead to interruptions in energy supply which however are not very dangerous since electricity consumption decreases during the night hours. Such interruptions are much less significant than irregular interruptions in operation of other power plants based on renewable energy resources (terrestrial solar, wind and tidal power plants).

The economic indices of SPS are expected to be similar to the indices of a Lunar power system and will be considered below.

Lunar power system (LPS). Three conceptions (varieties) of LPS will be considered below:

1. An initial LPS conception with light-reflecting mirrors in lunar orbits (LO) and satellites-retransmitters of SHF-rays in the orbits around the Earth (the Earth orbits — EO) which is suggested in [135]).

2. An LPS conception with additional basis of solar collectors on the back (invisible from the Earth) side of the Moon instead of mirrors in LO [138].

3. A "simplified" LPS conception without mirrors in LO and reflectors in EO [138].

Conceptions 1 and 2 provide uninterrupted power supply to the Earth (excluding the periods of the total lunar eclipses) and conception 3 provides interrupted power supply only at the periods when the Moon is seen from the point of the rectenna's location.

Figure 5.10 presents the initial (the first) LPS conception. Several pairs of bases with solar collectors and SHF-antennas are constructed on the Moon. Where possible they are located closer to the Moon's perimeter for at least one of them to be lighted as long as possible. However, owing to the fact that the side of the Moon visible from the Earth appears monthly entirely in shadow, a series of satellites with sunlight reflectors are launched around the Moon to light collectors during the eclipses.

MOON

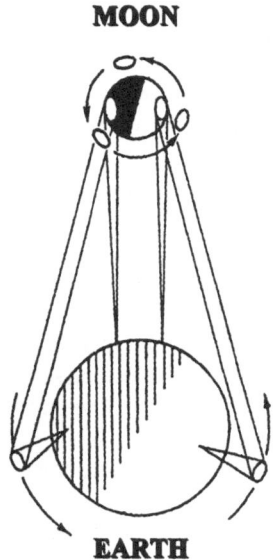

EARTH

Figure 5.10. Block diagram of Lunar power system.

Receiving rectennas are constructed on the Earth and satellites-retransmitters of SHF-radiation are launched. Rectennas receive power either directly from the Moon or via satellites-retransmitters. From the viewpoint of the authors, such an LPS scheme will provide an uninterrupted energy supply to the Earth.

The authors suppose that LPS will reach the capacity of 20 TW (received on the Earth) during a 40-year period of its development (construction) and will be in operation for the subsequent 30 years. Here based on their calculations the specific capital investments will make up US$ 400/kW and electricity cost US$ 0.002/kWh.

The project is attractive due to its huge capacity and very good specific economic indices. The latter are due to relatively low expenses on the space (lunar and orbital) part of the LPS — they make up only 13% of the total expenses. This is caused by maximum mechanised and robotised production of lunar materials and LPS elements on the Moon. It is supposed that the personnel staying on the Moon and in the orbits will make up about 5 thousand people all in all.

The "Earth" costs of the space part of the LPS will consist of research and development works, launch of some minimum of materials and equipment required to construct manned bases and startup production, salary of the personnel and expenses on its periodical replacement. All the rest, including preparation of mechanisms-robots and all possible constructions, erection of LPS elements, fuel production for rockets, will be carried out directly on the Moon, say, "free of charge" for the Earth.

The main disadvantage of this LPS conception is unreality of providing permanent lighting for lunar bases of collectors using mirrors in lunar orbits. These mirrors should permanently rotate directing the "reflected sun spot" to the collector. Based on the estimations of specialists in the sphere of astronautics, the maximum possible size of such a mirror will be about 1 km in diameter. Taking into account the area of solar collectors (tens of thousands of km^2) and a required number of mirrors per orbit for permanent lighting of one and the same collector site, about 1 million mirrors will be required in LO (at the LPS capacity of 20 TW). This seems to be absolutely impossible, even if only from the viewpoint of near-Moon space pollution.

The second conception of LPS with construction of additional solar collectors on the back side of the Moon looks more realistic. In fact this means their construction on three bases of which at least one is permanently lighted by the Sun (except for the total lunar eclipses). This increases the required area of solar collectors (photovoltaic cells) and necessitates constructing transmission lines from the bases on the back side of the Moon to the transmitting antennas. However these additional works on the Moon are much less labor intensive than installation and launch of mirrors to lunar orbits (with their further removal from there).

At the same time in this conception SHF-retransmitters are kept in orbits around the Earth, which requires special consideration. In principle, there are two possible options of their location: in geostationary and medium-height polar orbits.

The first method seems to be quite practical, though it requires special study of possible schemes and algorithms of switching SHF-rays from different antennas on the Moon to the satellites-retransmitters or directly to the receiving rectennas. Taking into account the stationary position of retransmitters in GSO, such schemes and algorithms will not be too complicated. However, in this case the constraints, similar to those considered above as applied to SPS, i.e. the constraints on the total number of retransmitters and latitudinal location of rectennas, will show up. The total capacity of LPS at such a scheme will not be very large (probably less than 1 TW). This type of LPS should be compared with Solar power satellites in terms of the implementation difficulties and economic efficiency (in particular, SPS of the same capacity will require a three times smaller area of photovoltaic cells).

Location of SHF-retransmitters in medium-height polar orbits casts serious doubts. This will necessitate tens of such satellites to service each rectenna. This is explained by the fact that the satellite flying over a rectenna will be visible only for a relatively short period of time and a series of many satellites in one and the same orbit is needed to provide continuous exposure of the rectenna. There will be a need for tens or even hundreds of thousands of retransmitters in Earth orbits depending on unit capacity and number of rectennas. This is unlikely to be technically and economically acceptable and allowable in terms of the near-Earth space pollution. Besides, an extremely complex control system would have been required to switch the SHF-rays from one satellite to the other or directly to the rectenna. Such a type of LPS should be either excluded from consideration or studied at its constrained parameters (in terms of the number of rectennas).

The third conception of LPS is the simplest and easiest to realise. It implies neither light reflectors in lunar orbits nor SHF-retransmitters in the orbits around the Earth (but as we will assume with additional bases of photovoltaic cells on the back side of the Moon). In this case the SHF-rays are transmitted from the lunar antennas directly to rectennas of the Earth. Thus, there will be long interruptions in power supply to rectennas. Besides, they can not be located in the polar zones of the Earth where the Moon does not rise high enough above the horizon. The same constraint is true for SPS as well.

This LPS conception supposes 14–18-hour interruptions in SHF-exposure of rectennas daily. This means the need for either back-up of LPS capacity by other kinds of power plants or use of energy storage installations. In the case of back-up, the power from Space will only save fuel (as would many other RES). Whereas the use of storage installations will require an increase in the capacity (and area) of rectennas (and lunar bases) 4–5 times compared to the average daily equaled capacity of LPS (with consideration for efficiency of storage installations). No doubt, this will make their power much more expensive.

To choose the best conception (or type) of LPS, additional comprehensive studies are required including comparison of LPS and SPS. The results of experimental works on the Moon development and creation of production facilities there will also be important.

Estimation of LPS economic indices. The supposition of the project's authors [135] on creating a Lunar power system with a capacity of 20 TW during 40 years seems to be unreal despite the above very good economic indices of LPS. Therefore these indices were estimated at a capacity of 2 TW and with consideration for uncertainty of some initial data [10].

Let us recall that the overwhelming majority of elements of the LPS space part, including solar collectors (photovoltaic cells), transformers and antennas of SHF-rays on the Moon, satellites-retransmitters around the Earth, etc. should be manufactured on the Moon from materials available there. This, as was already mentioned before, makes the project much cheaper — its cost is determined by the expenditures on the Earth. They are made up of: 1) expenses of construction and maintenance of rectennas and inverters on the Earth; 2) expenses of cargo transportation to orbits around the Earth and to the Moon; 3) cost of the transported cargoes; 4) salary of the personnel working in Space; 5) expenses of the R&D works. The following materials are delivered from the Earth to the Moon: some "initial" materials and equipment for creation of inhabited bases and necessary productions; materials and devices, whose manufacturing on the Moon is impossible; personnel working there (there and back) and the cargoes for their survival.

Technical and economic characteristics of LPS equipment elements and space transport systems are assumed based on the publications [137–140] and unpublished works kindly put at our disposal by D. Kriswell. The calculations were performed in 1990 US$ at the following preconditions.

The assumed length of an SHF-transmission wave is 12.24 cm (2.45 GHz) which provides its practical independence of weather conditions on the Earth. SHF-radiation density should be low enough to provide safety of people and nature and preservation of the Earth's ionosphere. There is great uncertainty on this point now — the indicated figures are from 10 to 500 W/m^2 (for comparison, solar radiation density of the Earth reaches 800–1000 W/m^2). In any case the area of rectennas on the Earth turns out to be rather large (see Table 5.15 with characteristics of two types of rectennas).

Table 5.15. Technical characteristics of rectennas

Characteristic	Rectenna "safe"	Rectenna "slightly dangerous"
Electric capacity (by DC), GW	1.0	2.0
Diameter, km	10.5	7.4
Area, km^2	87	43
Peak capacity of SHF-radiation (in the center of rectenna), W/m^2	60	240
Average capacity of SHF-radiation, W/m^2	12.8	51.7

It should be noted that a permissible capacity of SHF-radiation according to the standards of different countries is no more than 10–50 W/ m^2 at short-term (up to 2–3 h/day) and no more than 1–10 W/ m^2 at long-term exposure. Hence, even "safe" rectennas should be located far from densely populated areas.

Two options were considered for each type of rectenna: "cheap" and "expensive" rectennas with specific capital investments of US$ 45 and 75 per m^2 and specific operating costs of US$ 1.45 and 2.20 per m^2 per year respectively.

Efficiency of LPS elements is assumed to be improved starting from the current level (Table 5.16).

Essential differences in the cost of space cargo transportation to the Moon and near-Moon orbit depending on the delivery speed were taken into account by considering two options: "cheap" transportation (slow delivery) and "expensive" transportation (fast delivery).

Capital investments and operation costs were determined for a 70-year period, including:

— 10 years for creation of production infrastructure on the Moon;

— 30 subsequent years for production and erection of energy equipment and beginning of its operation (with the rate of 67 GW/year);

— 30 years for normal operation with replacement of equipment with an expired life span.

Table 5.16. Efficiency of SHF-transmission elements

LPS elements	Level	
	current	perspective
DC-to-SHF-radiation converters	0.7	0.9
Transmitting antenna	0.95	0.99
Transmission from the Moon to the Earth atmosphere	0.996	0.996
Transmission via the Earth atmosphere (depending on weather conditions)	0.92–0.98	0.92–0.98
Reflectors on the Earth orbit	0.9	0.98
Rectenna	0.85	0.93
Inverter	0.94	0.96
Total	0.45	0.74

Costs of cargo delivery and construction of rectennas and inverters on the Earth were discounted with the rate of 0.1. It was supposed that 60t of cargo per 1 GW of LPS capacity has to be launched into space; average cost of the cargoes is US$ 500/kg. The salary of personnel operating in Space is US$ 1.2 million /year.

Electricity cost was determined for two LPS conceptions:

1) "simplified" (LPS–1) that supposes only direct power transmission from the Moon surface to the Earth (without satellites around the Moon and the Earth but with additional bases of collectors on the back side of the Moon); the installed capacity of rectennas is used for about 2800 h/year;

2) "complete" (LPS–2) with additional bases on the Moon and SHF-reflectors in the orbits around the Earth, the installed capacity of rectennas is used during 5500 h/year.

Calculation of technical and economic indices for these LPS conceptions at different types and costs of rectennas is presented in Table 5.17. The space parts of LPS–1 and LPS–2 differ very little, therefore specific capital investments and operation expenses for both conceptions are assumed equal. The electricity costs are different due to different numbers of rectennas' exposure hours.

Table 5.17. Technical and economic indices of LPS

Indices	Type of rectenna			
	safe		slightly dangerous	
	cheap	expensive	cheap	expensive
Installed capacity on the Earth (by DC), GW	2000	2000	2000	2000
Receiving electric power on the Earth for 60 years (AC, LPS–2), 10^{12} kWh	556	556	556	556
Personnel in Space, people	440	440	440	440
Economic indices (for 70 years), 10^9 US$				
Capital investments in rectennas				
	7758	12930	1921	3201
Operating costs of rectennas	10326	15667	2557	3879
Capital investments in inverters	331	331	331	331
Operating costs of inverters	242	242	242	242
Costs of cargo transportation	283*	751**	283*	751**
Cargo cost	79	79	79	79
Salary of permanent staff	37	37	37	37
R&D works	100	100	100	100
Total	19156	30137	5550	8620
Extraordinary costs (15%)	2873	4521	833	1293
Total	22029	34658	6383	9913
Specific indices (per 1 kW of AC)				
Capital investments, US$/kW	5815	9459	1641	2567
Operating costs US$/kW/yr	83.6	127.4	34.3	37.9
Electricity cost (AC, cent/kWh)				
LPS–1	4.7	7.4	1.7	2.1
LPS–2	2.1	3.4	0.8	1.0

* "Cheap" transportation,
** "Expensive" transportation.

As is seen the obtained economic indices of LPS, on the one hand, vary greatly (in capital investments from US$ 1640 to 9460 per kW, in cost from 0.8 to 7.4 cent/kWh) due to uncertainty of its construction conditions and, on the other hand, they exceed greatly the estimates of the authors (US$ 400/kW and 0.2 cent/kWh). At the same time for a "slightly dangerous" rectenna these indices turned out to be quite good, close to the economic indices of the modern nuclear power plants.

In the calculations on the GEM-10R model it is assumed that the economic indices of LPS equal the values from the middle of their uncertainty ranges and get improved by the end of the 21st century (Table 5.18).

The specific economic indices of Solar power satellites, as it can be expected now, should be close to the LPS indices since the cost of terrestrial part (rectennas and inverters) is the same for LPS and SPS of equal capacity and differences in the space part affects insignificantly the common specific costs. Therefore it can be supposed that the model GEM-10R takes into account Space power systems as a whole (both LPS and SPS).

Table 5.18. Technical and economic characteristics of LPS

LPS	Year	Efficiency (net)	h, h/year	K, US$/kW	S, (US$/year)/ kW	C_{el}, cent/kWh	C_{mod}, US$/GJ (p.i.)
Simple	2025	0.74	2800	5500	220	21.06	43.28
	2050	0.77	2800	5000	200	19.14	40.94
	2075	0.80	2800	4500	180	17.23	38.28
	2100	0.83	2800	4000	160	15.31	35.31
Complete	2025	0.74	5500	6000	240	11.69	24.04
	2050	0.77	5500	5500	220	10.72	22.93
	2075	0.80	5500	5000	200	9.75	21.66
	2100	0.83	5500	4500	180	8.77	20.22

Part II.

STUDY ON PROBLEMS AND TENDENCIES OF ENERGY DEVELOPMENT IN THE 21st CENTURY

Chapter 6

GLOBAL SCENARIOS OF EXTERNAL CONDITIONS FOR ENERGY DEVELOPMENT

As was pointed out in the Preface, during long-term studies several series of calculations were performed on the model GEM-10R for a rather large number of scenarios of external conditions. Partially they were described in the papers [12–15]. The results of the calculations will be analysed for the eight global scenarios below, which seem to be the most interesting.

Three main factors varied in the considered scenarios. Their uncertainty affects greatly energy development of the world as a whole and in its regions in the 21st century:

— levels of final energy consumption;
— global constraints on CO_2 emissions;
— constraints on nuclear energy development.

Energy consumption levels determine general scales of energy development, consumption of non-renewable energy resources, energy costs, volumes of emissions, etc. In turn they depend on the rates and level of social and economic development of the regions and are, therefore, one of the most important characteristics of energy for sustainable development.

In Section 3.4 the demands for final forms of energy (electrical, thermal, mechanical, chemical) in the 21st century are forecasted for 10 world regions for high and low levels. The latter is performed assuming that the developed countries (the regions of North America, Europe, Japan and Republic of Korea, Australia and New Zealand) will render economic assistance to some developing regions (Africa, South and Southeast Asia, the Middle East) as well as to the former USSR. These two energy consumption levels are assumed in the corresponding scenarios.

The need to introduce *constraints on CO_2* emissions and their specific magnitude remains rather uncertain. There are different opinions, very often opposing. Documents from the Intergovernmental Panel on Climate Change (IPCC) [59] present a rather wide spectrum of forecasts on CO_2 emissions for the 21st century. No specific recommendations have been elaborated so far. In this connection quite a large range of such constraints were considered in the described studies starting with the case of their absence. Three variants of constraints on CO_2 emissions were assumed for the world as a whole – "rigid", "moderate" and "soft". They were supposed to change in time (Table 6.1).

Rigid constraints are assumed according to [68]. They suppose some increase in the magnitude of emissions in the mid-21st century with its subsequent decline by the end of the century to about the 1990 level. The temporary rise is admissible due to expected difficulties of fast reformation in the energy structure. It is believed that the scenario reflects realisation of the worst fears concerning impact of CO_2 emissions on the planet climate (its warming).

Table 6.1. Characteristics of global constraints on CO_2 emissions, Gt CO_2/year

Constraints	1990	2025	2050	2075	2100
Rigid	22	30	28	26	24
Moderate	22	35	40	40	40
Soft	22	37	45	50	55

Moderate constraints suppose that the magnitude of emissions will reach about 11 Gt C yearly by the end of the century, which, according to [69], will cause warming of the climate that still will not lead to catastrophic consequences in the next centuries, i.e. can be considered admissible. This variant was introduced after the calculations for rigid scenarios had shown the need for extremely great and unreal changes in the energy structure of most regions to meet such constraints (see Chapter 7).

We decided to consider scenarios with soft constraints on CO_2 emissions, after calculations on moderate constraints had shown that even under such conditions the projected energy structure of some regions would be hard to reach and would require large economic expenses, particularly with simultaneous constraints on nuclear energy development (see Section 7.1). Soft constraints should be considered as those corresponding to optimistic views on impact of anthropogenic CO_2 emissions on the global C cycle and on the planet climate.

Input of only global constraints on CO_2 emissions in the model GEM-10R (the model allows their introduction for individual regions as well) leads to distribution of these emissions among the regions almost proportionally to their fossil fuel consumption. Taking into account a considerable difference that still remains in the specific (per capita) energy consumption of the regions in the developed and developing countries, such a distribution of emissions (and costs related to their decrease) can not be considered quite "fair". Distribution of emissions proportionally to the population number in the regions, for instance, would have been more logical. This problem, associated with the concept of sustainable development of the world community, will specially be considered in Chapter 9.

It should be pointed out that the economic efficiency of *nuclear energy* for preventing planet climate change is evident compared to renewable energy sources, however there are well-known problems in providing proper safety of nuclear reactors and of the whole nuclear fuel cycle that caused the necessity to introduce *constraints on nuclear energy development*. The described studies consider two variants of such constraints – rigid, supposing introduction of a moratorium on nuclear energy development starting from 2025, and moderate, allowing its use to limited degrees in the 21st century.

Moderate constraints on nuclear energy use were formed in the following way. Since the preparation required for introduction of nuclear power plants (estimation of potentialities, site choices, feasibility study, design, construction) take quite a long period of time, one may consider that for the nearest 10–15 years nuclear energy development is predetermined by the current state and existing inertia. Therefore we used the forecasts to 2010–2015 made by the experts of IAEA based on national nuclear energy programs which were then extrapolated up to 2025 (at the same development rate).

Presently nuclear energy is used on an industrial scale in 32 countries of the world mainly for production of electric power. In 18 countries the fraction of electric power produced by nuclear power plants (NPP) exceeds 20%, in eight countries – 40% and in three (France, Belgium, Lithuania) – 50%. Proceeding from the above and from the fact that the governments of many countries form energy programs based on the need to diversify energy supply sources, nuclear energy use for electric power production was limited by the 50% share in each region of the world after 2025. This share was distributed among the nuclear power plants with thermal and fast breeder reactors as follows (numerator – thermal, denominator – fast reactors): 50/0 in 2025, 30/20 in 2050, 20/30 in 2075, 10/40 in 2100. In the regions of the Middle East and Africa introduction of NPP with fast reactors in 2050 was excluded.

The measures of nuclear power use for other purposes (e.g. production of hydrogen and heat) were limited in each region to the magnitude of its use for electric power production.

Table 6.2 presents the variants of the above three main uncertain factors assumed in the eight Scenarios. Notations of the scenarios, which will be used hereinafter along with their numbers, will contain the first letters in the names of the corresponding variants of conditions (factors). In doing so the sequence of letters corresponds rigidly to the order in which the three factors are considered above and located in Table 6.2: variants of energy consumption levels, constraints on CO_2 emissions and on nuclear energy development. For instance Scenario 1 is denoted as HNN (high, no constraints, no constraints) and Scenario 5 HRM (high, rigid, moderate).

Table 6.2. Scenarios of external conditions of energy development

Factor	Numbers and notation of scenarios							
	1 HNN	2 HNR	3 HRN	4 HRR	5 HRM	6 HMN	7 LMM	8 LSM
Energy consumption level	High	High	High	High	High	High	Low	Low
Constraints on CO_2 emissions	No	No	Rigid	Rigid	Rigid	Moderate	Moderate	Soft
Constraints on nuclear energy development	No	Rigid	No	Rigid	Moderate	No	Moderate	Moderate

As was pointed out the scenarios were formed and selected as runs on the model GEM-10R were performed and the obtained results were analysed. First, the calculations were performed for the first four Scenarios with rigid constraints (or without them). Then Scenarios 5 and 6 were formed and calculated for the same high level of energy consumption but with partial transition to moderate constraints. Analysis of their results has shown that the obtained variants of the energy development of the world and its regions are still hard to realise and do not quite completely meet the requirements and conditions of world sustainable development in social-economic aspects. In particular, the considered energy consumption level, on the one hand, was hard to reach and, on the other hand, did not ensure a desirable approach of developing countries to the developed ones in the level of per capita energy consumption and economic development (specific GDP).

In this connection a new forecast of energy consumption was made that assumed a faster decrease in energy-GDP ratio and economic assistance of the developed countries to the developing ones (see Section 3.4). For the obtained lower energy consumption, first Scenario 7 and then Scenario 8 with soft constraints on CO_2 emissions were calculated. Scenario 8 is considered by the authors as the best to meet the requirements of sustainable development. The authors venture to call energy development of the regions and the world as a whole obtained for this variant "plausible" and to consider it as quite a probable forecast.

In general the following points (circumstances) were taken into account when choosing scenarios of external conditions for energy development:

1. *Comparability of the variants* for estimating the consequences of constraints. This concerns Scenarios 1–4, the calculations for which had to estimate the impact of a nuclear moratorium and rigid constraints on CO_2 emissions on the energy structure and related economic expenditures. The results obtained for these scenarios and compared pair-wise allow one to determine the "prices" of a moratorium on nuclear energy development and of constraints on CO_2 emissions.

2. *Determination of CO_2 emission levels* that can be considered acceptable (desirable) in terms of "normal" energy development, i.e. a sort of requirements of the energy itself to the magnitude of constraints on these emissions. This objective has affected greatly the content (structure) of Scenarios 6–8.

3. *Search for trade-off (realistic) decisions on nuclear energy development.* This concerns Scenarios 5–8. This objective, obviously, can not be considered as completely attained since only one variant of such constraints was considered, however a definite step in this direction has been made.

4. Analysis of the conditions and requirements of the world *sustainable development*. Mainly this concerns Scenarios 7 and 8.

In the next chapters the results of the calculations on the model GEM-10R for the eight considered Scenarios will be analysed in detail.

Chapter 7

CHANGES IN THE WORLD ENERGY STRUCTURE

7.1. Primary energy consumption

The results of calculations on the model GEM-10R for all eight Scenarios are presented in Figure 7.1. Recall the assumption made: a higher level of energy consumption for Scenarios 1–6; a nuclear moratorium for Scenarios 2 and 4; constraints of CO_2 emissions that are rigid for Scenarios 3, 4, and 5, moderate for Scenarios 6 and 7, and soft for Scenario 8. Moderate constraints on nuclear energy (NE) development are introduced in Scenarios 5, 7 and 8.

The results obtained for Scenarios 1–4 can be considered as extreme and their realisation is unlikely:

— in Scenarios 1 and 2 mostly coal is used — its fraction reaches 50–60% in 2100;

— Scenario 3 can be named "nuclear"— the nuclear energy fraction exceeds 70% in 2100;

— in Scenario 4 renewables (biomass and non-fossil renewables (NFR)) are prevailing (more than 50% in 2100).

In coal Scenarios 1 and 2 with no constraints on CO_2 emissions, the latter appeared to be very large — about four times greater by the turn of the 21st century than in 1990 (Tables 7.1 and 7.2). Such CO_2 emissions can hardly be recognised as admissible, the more so as they are simultaneously accompanied by the same substantial (3–5 times) increase in emissions of sulfur oxide, nitrogen oxide and particulates (Table 7.3).

Table 7.1. CO_2 emissions by the world energy, billion t/year

Year	Scenario							
	1 HNN	2 HNR	3 HRN	4 HRR	5 HRM	6 HMN	7 LMM	8 LSM
1990	22	22	22	22	22	22	22	22
2025	46	46	30	30	30	33	33	37
2050	59	67	28	28	28	40	40	45
2075	80	84	26	26	26	40	40	50
2100	90	104	24	24	24	40	40	55

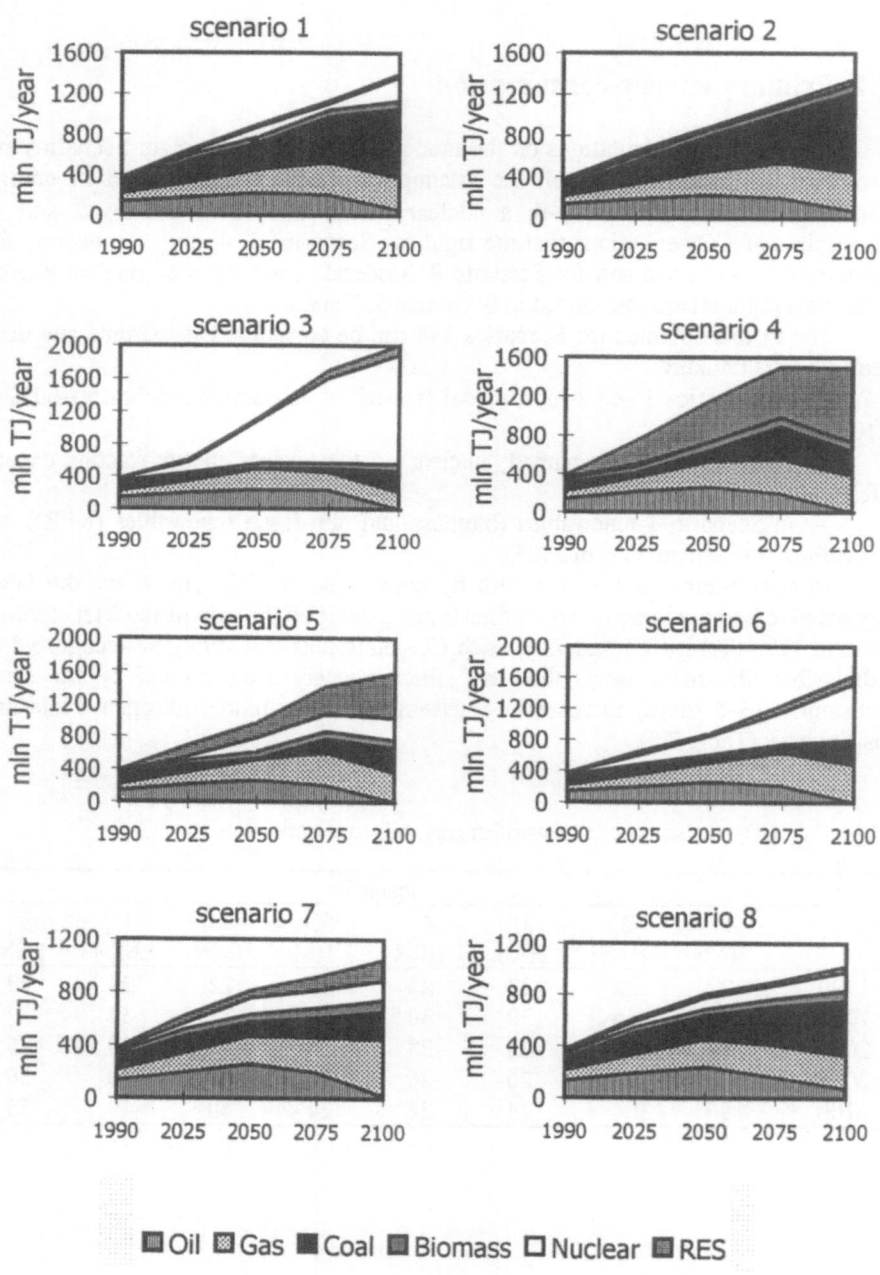

Figure 7.1. Primary energy consumption (Scenarios 1–8).

Table 7.2. Volumes of CO_2 removal, billion t/year

Year	Scenario							
	1 HNN	2 HNR	3 HRN	4 HRR	5 HRM	6 HMN	7 LMM	8 LSM
1990	0	0	0	0	0	0	0	0
2025	0	0	1.3	1.5	0	0	0	0
2050	0	0	1.7	14.2	3.6	0	0	0
2075	0	0	4.4	39.0	23.1	0	0	0
2100	0	0	9.2	20.2	19.6	0	0	0

Scenario 3 (as well as Scenario 6, Figure 7.1) shows that nuclear energy is the most economic for decreasing CO_2 emissions. However the optimal scales of nuclear energy development obtained on the model for Scenario 3 seem to be absolutely unrealistic, the more so that starting in 2050 nuclear energy development should be based on fast reactors (breeders) that will use "waste" uranium-238 (Table 7.4). The resources of natural uranium-235 are depleted already by 2050 (this will be considered in more detail in the next section).

Table 7.3. Emissions of harmful substances, million t/year

Year	Scenario							
	1 HNN	2 HNR	3 HRN	4 HRR	5 HRM	6 HMN	7 LMM	8 LSM
				SO_2				
1990	113.6	113.6	113.6	113.6	113.6	113.6	113.6	113.6
2025	147.1	156.3	76.6	75.9	83.2	97.1	110.2	134.9
2050	166.8	212.5	40.3	31.3	45.3	78.7	101.4	147.1
2075	220.7	232.1	34.6	35.3	31.4	33.1	89.7	136.2
2100	206.4	290.7	19.9	14.0	18.0	20.0	34.5	149.9
				NO_x				
1990	74.6	74.6	74.6	74.6	74.6	74.6	74.6	74.6
2025	157.0	160.7	107.9	111.5	120.3	120.7	134.0	147.6
2050	201.1	231.4	142.4	138.6	141.2	145.4	166.7	187.2
2075	274.3	284.7	164.0	160.9	165.1	175.3	171.8	202.4
2100	264.2	332.6	174.5	178.1	183.4	199.7	159.5	205.5
				Particulates				
1990	63.3	63.3	63.3	63.3	63.3	63.3	63.3	63.3
2025	143.0	152.7	36.4	34.1	48.5	64.5	92.3	127.1
2050	160.1	225.7	13.5	14.0	21.1	36.8	85.4	140.7
2075	236.0	266.8	18.5	20.8	23.3	19.3	80.7	140.4
2100	192.9	331.1	24.2	16.0	26.5	24.9	31.2	136.2

Table 7.4. Application of thermal (U-235) and fast reactors (U-238) and overall scales of the nuclear energy, million TJ/year

Year	Scenario							
	1	2	3	4	5	6	7	8
	HNN	HNR	HRN	HRR	HRM	HMN	LMM	LSM
				U-235				
1990	23.2	23.2	23.2	23.2	23.2	23.2	23.2	23.2
2025	34.4	23.0	127.8	24.0	44.0	111.6	43.8	43.4
2050	101.8	0.0	28.1	0.0	85.3	28.1	80.1	80.2
2075	54.8	0.0	43.2	0.0	42.5	41.1	48.5	46.9
2100	31.2	0.0	23.0	0.0	27.0	18.0	26.4	28.3
				U-238				
1990	0.0	0.0	0.0	0.0	0.0	0.0	0.0	0.0
2025	0.0	0.0	0.0	0.0	0.0	0.0	0.0	0.0
2050	0.0	0.0	512.7	0.0	80.9	215.0	39.0	6.2
2075	5.5	0.0	998.3	0.0	169.4	378.6	84.7	58.2
2100	205.0	0.0	1312.4	0.0	239.3	703.8	119.7	119.7
				Total				
1990	23.2	23.2	23.2	23.2	23.2	23.2	23.2	23.2
2025	34.4	23.0	127.8	24.0	44.0	111.6	43.8	43.4
2050	101.8	0.0	540.8	0.0	166.2	243.1	119.1	86.3
2075	60.3	0.0	1041.5	0.0	211.9	419.7	133.2	105.3
2100	236.2	0.0	1335.4	0.0	266.3	721.8	146.1	148.1

In Scenario 4 renewable energy sources should expand on a vast scale (see Figure 7.1) due to rigid constraints on CO_2 emissions and a nuclear moratorium. In 2025 their fraction should exceed 30% and the scales of application should increase more than 25 times compared to 1990. Here a considerable portion of energy should be received from the Lunar Power System — LPS (Table 7.5). Obviously, this Scenario is absolutely unrealistic for the above reasons as well as for the expenses of energy development (Figure 7.2).

Based on analysis of the expenses it is possible to determine the "price" of introducing a nuclear moratorium and constraints on CO_2 emissions. Such an analysis can be carried out for Scenarios 1–6 with the same level of energy consumption for 2100 (as is seen in Figure 7.2 the proportions of expenses turn out to be approximately the same for 2050 and 2075). These expenses are given in Table 7.6.

Comparison of the expenses for Scenarios 1 and 2 shows that introduction of a moratorium on nuclear energy development increases them by 2% only (in Figure 7.2 they are indistinguishable). This is because nuclear and coal power plants are approximately equally economic, and replacement of the former by the latter (in the corresponding regions) affects expenses insignificantly (but, as was noted, it results in a considerable increase of harmful emissions — see Tables 7.1 and 7.3).

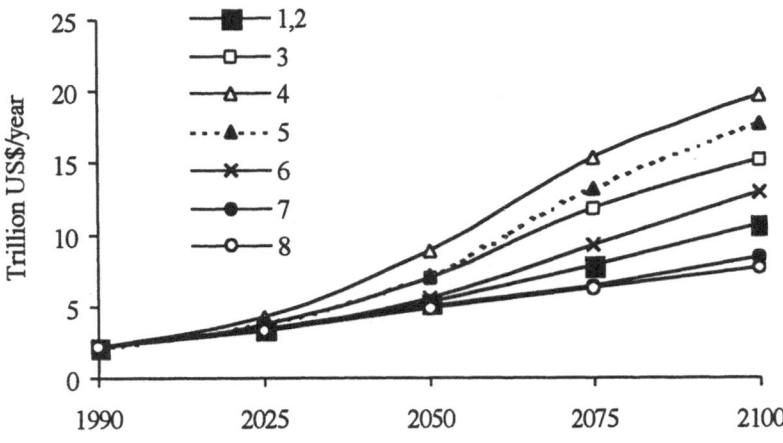

Figure 7.2. Expenses of world energy in Scenarios 1–8.

Introduction of constraints (rigid) on CO_2 emissions increases the expenses substantially — by 42% in Scenario 3 and by 85% in Scenario 4, under conditions of nuclear moratorium. Hence, simultaneous introduction of both constraints practically doubles the energy development costs. The "fraction" of moratorium in these costs amounts to about a half (43% or 4.6 trillion US$/year as a difference between Scenarios 4 and 3).

Mitigation of constraints on the nuclear energy development (Scenario 5) decreases costs but insignificantly (by approximately 20%). Here the impact of constraints on CO_2 emissions remains great.

Scenario 6 differs from Scenario 3 by moderate constraints on CO_2 emissions (see Table 7.1). In this connection in Scenario 6 the fossil fuel fraction increases, particularly the gas fraction (Table 7.7) and the nuclear energy fraction decreases. However the scales of the nuclear energy development remain oversize (see Figure 7.1). Furthermore even under moderate constraints on CO_2 emissions, coal is almost completely displaced from the energy balance of the 21st century (as in Scenario 3).

Particularly this concerns electric power generation (Table 7.8). It can be seen that in Scenario 6 coal is used at power plants only till 2025 and on a scale smaller than in 1990 (in Scenario 3, coal is inadmissible for electric power production even in 2025). Such a situation should also be considered unacceptable particularly for such regions as China, the former USSR, South and South-Eastern Asia.

Table 7.5. Use of non-fossil renewables, million TJ/year

Year	Scenario							
	1 HNN	2 HNR	3 HRN	4 HRR	5 HRM	6 HMN	7 LMM	8 LSM
	Hydropower							
1990	9.3	9.3	9.3	9.3	9.3	9.3	9.3	9.3
2025	21.2	21.2	21.2	32.8	30.1	21.2	28.2	21.6
2050	21.2	21.2	21.2	21.2	24.0	23.0	23.5	21.3
2075	21.2	24.3	21.2	21.2	21.2	21.2	21.7	22.3
2100	21.2	26.5	21.2	21.2	21.2	21.2	21.2	27.7
	Wind energy							
1990	0.1	0.1	0.1	0.1	0.1	0.1	0.1	0.1
2025	0.1	1.3	9.0	102.9	75.8	9.0	26.2	0.3
2050	2.7	3.3	9.0	110.5	85.6	9.0	35.4	9.0
2075	9.0	9.0	9.0	114.8	113.1	9.0	66.0	9.0
2100	9.0	20.7	9.0	112.5	113.0	9.0	80.6	17.9
	Solar energy							
1990	0.0	0.0	0.0	0.0	0.0	0.0	0.0	0.0
2025	0.0	0.0	0.0	67.3	23.2	0.0	0.0	0.0
2050	0.0	0.0	0.0	171.8	27.0	0.0	0.0	0.0
2075	0.0	0.0	62.7	183.5	105.8	61.3	14.5	0.0
2100	0.0	0.0	90.7	171.8	128.2	81.2	35.7	0.0
	Geothermal energy							
1990	0.0	0.0	0.0	0.0	0.0	0.0	0.0	0.0
2025	0.0	0.0	0.0	37.3	12.7	0.0	0.0	0.0
2050	0.0	0.0	0.0	42.3	22.0	0.0	0.0	0.0
2075	0.0	0.0	0.0	47.1	32.2	0.0	0.0	0.0
2100	0.0	0.0	0.0	14.0	14.0	0.0	4.8	0.0
	LPS							
1990	0.0	0.0	0.0	0.0	0.0	0.0	0.0	0.0
2025	0.0	0.0	0.0	5.5	0.0	0.0	0.0	0.0
2050	0.0	0.0	0.0	71.2	24.9	0.0	0.0	0.0
2075	0.0	0.0	0.0	147.6	67.4	0.0	0.0	0.0
2100	0.0	0.0	0.0	445.0	265.4	0.0	16.1	0.0
	Total							
1990	9.4	9.4	9.4	9.4	9.4	9.4	9.4	9.4
2025	21.3	22.5	30.2	246.0	141.8	30.2	54.4	21.9
2050	23.9	24.5	30.2	417.0	183.3	32.0	58.8	30.3
2075	30.2	33.3	92.9	514.3	339.3	91.5	102.2	31.3
2100	30.2	47.2	120.9	764.5	541.5	111.4	158.4	45.6

Table 7.6. World energy costs in 2100

Costs	Scenario					
	1 HNN	2 HNR	3 HRN	4 HRR	5 HRM	6 HMN
Trillion US$/year	10.7	10.9	15.2	19.8	17.8	13.2
%	100.0	101.9	142.1	185.0	166.4	129.4

Table 7.7. Fossil fuel production, million TJ/year

Year	Scenario							
	1 HNN	2 HNR	3 HRN	4 HRR	5 HRM	6 HMN	7 LMM	8 LSM
	Oil							
1990	133.2	133.2	133.2	133.2	133.2	133.2	133.2	133.2
2025	178.0	183.0	199.5	211.3	204.6	197.4	193.5	189.4
2050	227.9	227.7	238.7	262.9	263.7	257.1	255.0	237.8
2075	202.6	199.0	225.2	189.9	195.6	208.8	196.3	162.4
2100	56.6	73.2	2.3	1.5	1.8	2.3	20.8	75.4
	Gas							
1990	77.2	77.2	77.2	77.2	77.2	77.2	77.2	77.2
2025	145.5	145.5	145.5	145.5	145.5	145.5	145.5	145.5
2050	219.6	219.6	255.5	259.6	242.1	245.8	210.8	208.7
2075	311.8	332.3	223.2	356.3	391.1	423.6	256.2	243.3
2100	315.1	372.4	210.5	378.4	349.2	470.5	419.2	252.5
	Coal							
1990	91.2	91.2	91.2	91.2	91.2	91.2	91.2	91.2
2025	223.4	223.4	58.6	52.3	83.2	98.8	124.3	170.4
2050	270.1	348.0	0.0	111.9	8.7	55.7	106.9	190.8
2075	458.7	478.9	48.2	343.0	172.5	70.2	147.8	300.6
2100	673.6	775.5	250.0	257.0	309.8	279.5	208.1	421.5
	Biomass							
1990	37.5	37.5	37.5	37.5	37.5	37.5	37.5	37.5
2025	47.1	47.3	44.4	65.8	65.8	44.4	54.6	42.9
2050	58.9	58.9	58.9	91.9	91.9	49.0	60.2	52.7
2075	69.7	69.9	69.9	91.9	91.9	28.2	77.7	60.1
2100	81.9	84.9	84.9	91.9	91.9	39.3	91.9	78.4
	Total							
1990	339.1	339.1	339.1	339.1	339.1	339.1	339.1	339.1
2025	594.0	599.2	448.1	475.0	499.4	486.0	518.0	548.3
2050	776.6	854.1	553.0	726.3	606.4	607.5	632.9	690.2
2075	1042.8	1080.1	566.5	981.1	851.2	730.9	677.9	766.4
2100	1127.4	1305.9	547.7	728.9	752.8	791.7	740.0	828.3

Table 7.8. Consumption of energy resources for electric power generation, million TJ/year

Year	Scenario							
	1	2	3	4	5	6	7	8
	HNN	HNR	HRN	HRR	HRM	HMN	LMM	LSM
Liquid fuels								
1990	13.3	13.3	13.3	13.3	13.3	13.3	13.3	13.3
2025	23.0	23.0	30.1	34.5	25.8	24.6	21.8	21.2
2050	31.6	31.9	46.1	0.0	17.9	34.2	29.7	29.6
2075	42.4	41.5	29.7	0.0	0.0	29.3	18.4	24.6
2100	40.8	34.3	0.3	0.0	0.0	0.3	10.8	32.1
Gas								
1990	14.2	14.2	14.2	14.2	14.2	14.2	14.2	14.2
2025	0.0	0.0	0.0	0.0	2.7	0.0	1.1	1.0
2050	0.0	0.0	35.6	0.0	17.1	5.6	4.2	1.6
2075	0.2	0.4	0.0	0.0	0.0	10.1	26.9	9.1
2100	9.2	13.0	3.1	0.0	0.0	84.3	45.8	6.1
Coal								
1990	50.5	50.5	50.5	50.5	50.5	50.5	50.5	50.5
2025	121.2	130.0	0.0	0.0	11.7	27.3	64.7	104.6
2050	135.0	210.7	0.0	0.0	0.0	0.0	65.2	119.6
2075	216.3	254.1	0.0	0.0	0.0	0.0	48.4	112.7
2100	122.3	293.6	0.0	0.0	0.0	0.0	0.0	82.1
Hydrogen								
2025	0.0	0.0	0.0	0.0	0.0	0.0	0.0	0.0
2050	0.0	0.0	0.0	48.9	12.1	0.0	0.0	0.0
2075	0.0	0.0	82.4	56.6	70.2	0.0	0.0	0.0
2100	0.0	0.0	123.6	141.8	133.4	16.2	0.5	0.0
Nuclear energy								
1990	23.2	23.2	23.2	23.2	23.2	23.2	23.2	23.2
2025	34.4	23.0	182.3	23.0	43.6	130.8	43.4	43.4
2050	101.8	0.0	529.1	0.0	111.3	243.1	119.1	86.3
2075	60.3	0.0	840.1	0.0	122.4	392.6	133.1	105.2
2100	236.2	0.0	1021.1	0.0	140.6	700.8	143.3	147.9
Renewables								
1990	10.2	10.2	10.2	10.2	10.2	10.2	10.2	10.2
2025	21.3	22.5	30.2	246.0	161.2	30.2	65.5	30.0
2050	23.9	24.5	21.2	276.0	183.3	32.0	65.2	30.3
2075	30.2	33.3	30.2	330.8	339.3	30.2	102.6	31.3
2100	30.2	47.2	30.2	592.7	541.5	30.2	160.5	45.6
Total								
1990	111.4	111.4	111.4	111.4	111.4	111.4	111.4	111.4
2025	199.9	198.4	242.6	303.5	245.1	212.8	196.5	192.1
2050	293.7	267.1	641.0	324.9	341.6	314.9	283.6	267.4
2075	349.3	329.2	982.4	387.4	532.0	480.1	329.5	282.9
2100	438.7	388.3	1178.3	734.5	815.5	831.9	360.9	313.8

Under moderate (Scenario 6) and rigid (Scenarios 3 and 4) constraints on CO_2 emissions of all the coal produced after 2025 are converted into synthetic fuel (see the next section). Here the technologies for CO_2 removal should be developed, particularly in Scenario 4 (see Table 7.2). The necessity of applying these technologies is one of the reasons causing a significant increase in energy costs in Scenarios 3–5.

The above circumstances and more thorough analysis of liquid fuel use for the needs of the transport and chemical industry (as performed in the following section) show that even moderate constraints on CO_2 emissions create an irrational energy structure in many regions of the world (at a high level of energy consumption). This relates to the excessively large scale required for nuclear energy development (or renewables), extremely small admissible volumes of coal utilization, large volumes of CO_2 removal and some other problems. Thus, further moderation of constraints on CO_2 emissions should be considered feasible. Such soft constraints are assumed in Scenario 8 (see Table 7.1).

The assumed level of final energy consumption is too high — this is one more conclusion that can be made based on the analysis of primary energy consumption in Scenarios 1–6. The total final energy consumption assumed for these Scenarios will increase almost four times by 2100 compared to 1990 (see Table 3.18 in Section 3.4). Primary energy consumption obtained for the Scenarios with constraints on CO_2 emissions (Scenarios 3–6) will also increase 3.5 – 4.7 times.

Taking into account uneven distribution of energy resources by the world regions, this means approximately a 10-fold increase in primary energy production in the fuel exporting regions (the Middle East, the former USSR, North America). Such an increase of energy production is hard to imagine. Therefore energy consumption assumed in Scenarios 1–6 should be considered as not-meeting the requirements and conditions of humanity's transition to sustainable development (it is spoken of in Chapters 3 and 9). Transition to a low level of energy consumption assumed for Scenarios 7 and 8, along with mitigation of constraints on CO_2 emissions, will considerably soften the requirements for energy and create possibilities to provide its more rational structure.

Use of primary energy resources in Scenarios 7 and 8 (see Figure 7.1.) turns out to be more uniform than in the other Scenarios. During most of the 21st century the share of all the main kinds of resources does not exceed 30% and increases to 40% for natural gas (Scenario 7) and coal (Scenario 8) only in the end of the century.

At a low level of energy consumption, moderate constraints on CO_2 emissions (Scenario 7) give quite a satisfactory structure of primary energy use. Compared to 1990, by the end of the 21st century the use of natural gas will increase 5.4 times and that of non-fossil renewables 17 times; this apparently is realisable.

Mitigation of constraints on CO_2 emissions in Scenario 8 leads, compared to Scenario 7, to an increase in total fossil fuel consumption (at the expense of decreased use of non-fossil renewables), to a considerable decrease in natural gas use and, to the contrary, to an increase in coal consumption (see two right-hand columns of Table 7.7). Use of electricity production (see Tables 7.5 and 7.8) decreases greatly and the need to use space energy (from LPS in 2100) loses significance. Expenses of energy (see Figure 7.2) decrease somewhat (by 10% in 2100) but the emissions of SO_2, NO_x and particulates (see Table 7.3) increase considerably.

Hence, the energy structure obtained for Scenario 8 is in some respects better — and in some worse — than for Scenario 7. This testifies to the fact that further mitigation of constraints on CO_2 emissions is not needed for low energy consumption. The level of CO_2 emissions (see Tables 6.1 and 7.1) assumed for soft constraints should be considered limiting in terms of energy structure feasibility for the world as a whole. It can even be supposed that the best variant of energy structure is somewhere in the middle between the variants for Scenarios 7 and 8. At the same time, on the whole the structure for Scenario 8 seems to be more preferable than for Scenario 7, i.e. more realistic scales of using coal and non-fossil renewables by power plants (see Table 7.8), "smoother" nuclear energy development (see Table 7.4), lesser expenses (see Figure 7.2), etc. Therefore Scenario 8 is chosen for a more detailed analysis (Sections 7.2–7.4 and Chapter 8).

Let us consider general trends in using primary energy resources in the 21st century that can be found from the analysis of the calculation results for all eight Scenarios (see Figure 7.1).

Use of *oil* in all Scenarios is almost similar — its production peak is reached in 2050 and cheap oil resources are depleted completely or almost completely by 2100 (this will be considered in more detail in the following section). The tendency is persistent owing to a great efficiency of oil products for production of mechanical and chemical energy as well as peak power and heat (see Section 7.4). In the end of the century oil will be replaced by synthetic fuel, first of all, from coal, which is more economical than expensive non-traditional oil (see Section 7.3).

Natural gas production increases permanently during the whole century reaching its maximum in the end of the century. Scenario 3 is an exception. Gas is used for production of all final energy forms but mostly for heat production (see Section 7.4). Partially it is converted into methanol and hydrogen. Natural gas production increases 4–6 times by the end of the century against 1990.

Coal and nuclear energy undergo the greatest changes depending on the constraints imposed. Being approximately equally economical they replace each other, particularly in the "extreme" Scenarios 2 and 3 (see Figure 7.1). They are used mostly at electric power plants. In the second half of the century a considerable part of coal will be converted into synthetic liquid fuel, mainly into hydrocarbons of light fractions. Nuclear energy in the Scenarios with rigid constraints on CO_2 emissions will be mostly used for hydrogen production (see Section 7.3), which replaces, in particular, liquid and gaseous fuels when producing peak power.

Use of *renewable energy sources* (see Tables 7.5 and 7.8) differ essentially in different Scenarios. Only hydropower, biomass and cheap wind resources are used permanently. The remaining kinds of renewable energy resources are the most expensive, they close the energy balance and are developed as needed. In Scenarios 4 and 5 the required application scale of renewables seems to be practically unreal. Efficiency of different renewables technologies is considered in more detail in Section 7.4.

7.2. Depletion of cheap fuel resources and trends in changes of their prices

Division of fuel reserves into several cost categories (up to eight) allows one, on the one hand, to more accurately follow the dynamics of their depletion (and mutual competition) and, on the other hand — to reveal general trends in changes of fuel prices in the 21st century. Both are of great interest when analysing long-term future of world energy development.

Fuel consumption of different cost categories, by time periods, obtained in Scenario 8 is presented in Table 7.9. Since the Scenario is most realistic in terms of sustainable development, the results obtained are very interesting, though similar data are obtained when calculating on GEM-10R model for all the other Scenarios as well. In Scenarios 1–6 with high energy consumption, fuel consumption is of course higher than in Scenario 8. Consider consumption of reserves for each fuel kind.

In 2025 *oil* resources of the first two categories are mainly used, though in some regions oil of the third category is used as well. One can suppose that world oil prices will be formed based on the cost of the most expensive category involved in the optimal solution of the model. Therefore there are good grounds to suppose that in 2025, world oil prices will be established based on the cost of the third category.

In 2050, oil of mainly the third and fourth categories is consumed. The remains of oil of the first two categories are produced in the region of the Middle East where oil reserves are the largest. It is safe to suppose that world oil prices in 2050 will correspond to the oil cost of the fourth category.

By 2075 the first three categories will be depleted in all regions. Production of oil of the fourth category ceases. It is natural to suppose that world oil prices in 2075 will also be formed based on the cost of the fourth category.

In the very end of the century, oil of the fifth category is consumed. Oil of the sixth and more so of the seventh and eighth categories, being "non-traditional" oil contained in bituminous sandstones, shales, etc., appears to be non-competitive in the 21st century compared to other energy resources. Hence, by 2100, world oil prices will correspond to its fifth cost category.

The situation with *natural gas* is more diversified, partially due to its difficult transportability. The largest gas reserves are concentrated in regions of the Middle East, the former USSR and North America, but are also found in the other regions.

Depending on the magnitude of resources of each category, their consumption in different regions ceases at different periods of the 21st century. Therefore at each period of time, gas of many cost categories is simultaneously produced in the world. Thus, gas of the first three categories is produced in 2025 and of four and even of five categories in the year 2050 and further.

Table 7.9. Production of fuels (world as a whole), million TJ/year (Scenario 8 — LSM)

Fuel type and category	Cost*, $/GJ	2025	2050	2075	2100
Oil -1	1.8	99.7	24.3	0.0	0.0
Oil -2	2.8	59.7	32.5	0.0	0.0
Oil -3	3.9	30.0	111.8	0.0	0.0
Oil -4	5.4	0.0	69.2	162.4	0.0
Oil -5	6.5	0.0	0.0	0.0	75.4
Oil -6	8.0	0.0	0.0	0.0	0.0
Oil -7	10.1	0.0	0.0	0.0	0.0
Oil -8	19.6	0.0	0.0	0.0	0.0
Gas-1	2.0	81.6	56.0	1.4	0.0
Gas -2	2.9	41.9	78.8	62.5	0.0
Gas -3	4.3	22.0	49.3	91.9	82.2
Gas -4	5.5	0.0	24.5	10.1	24.7
Gas -5	6.4	0.0	0.1	77.5	92.4
Gas -6	7.6	0.0	0.0	0.0	53.3
Gas -7	9.0	0.0	0.0	0.0	0.0
Gas -8	18.1	0.0	0.0	0.0	0.0
Coal-1	1.1	118.2	17.9	0.0	0.0
Coal -2	1.6	49.5	110.6	217.2	148.9
Coal -3	2.0	2.7	62.3	83.4	109.0
Coal -4	2.3	0.0	0.0	0.0	147.2
Coal -5	2.8	0.0	0.0	0.0	16.4
Uranium-235-1	0.5	4.3	0.0	0.0	0.0
Uranium-235-2	0.6	29.2	0.7	0.0	0.0
Uranium -235-3	0.7	9.9	70.2	6.6	0.0
Uranium -235-4	0.8	0.0	3.4	8.3	0.0
Uranium -235-5	1.0	0.0	2.7	10.9	0.0
Uranium-235-6	1.2	0.0	3.1	12.4	0.0
Uranium -235-7	1.4	0.0	0.0	7.4	10.2
Uranium -235-8	1.6	0.0	0.0	1.4	18.1
Uranium -238	0.5	0.0	6.2	58.2	119.7
Biomass-1	1.9	36.7	36.6	39.3	39.3
Biomass -2	2.6	4.8	11.9	13.5	30.9
Biomass -3	3.5	1.4	4.2	7.3	8.2
Total including:		591.6	776.4	871.6	975.7
oil	1.8–6.5	189.4	237.8	162.4	75.4
gas	2.0–7.6	145.5	208.7	243.3	252.5
coal	1.1–2.8	170.4	190.8	300.6	421.5
uranium	0.5–1.6	43.4	86.4	105.1	148.0
biomass	1.9–3.5	42.9	52.7	60.1	78.4

* In situ.

Gas of the first category is depleted practically only by the year 2050, gas of the second category — by 2100. Production of the third gas category extends through the whole 21st century. Use of the fourth and fifth categories starts in 2050 and that of the sixth — after 2075. Gas of the seventh and eighth categories (representing "non-traditional" gas and gas hydrates) appears non-competitive through the whole 21st century.

The prices of piped natural gas will be formed individually for the corresponding regions. The prices of liquefied natural gas can be set up for the world as a whole but at a higher level. The forecast on prices for piped gas is presented below for the region of Europe and for liquefied gas for the region of Japan and Republic of Korea.

Coal consumption grows considerably in the given Scenario. Its first three categories are used up to 2050, when the first category will be depleted. In the second half of the century the second and the third categories are extracted, and in the end of the century the fourth and the fifth are produced as well. The sixth and the next categories are not used in the 21st century (they are not presented in Table 7.9).

It can be expected that world coal prices will be formed based on the costs of coal production in the exporting regions (NA, AZ, SU). In the rest of the regions more expensive coal categories will be used taking into account bad transportability of coal and energy supply peculiarities in these regions. Therefore an additional analysis of cost categories of coal produced in different world regions was performed to forecast world coal prices.

The situation with consumption of *natural uranium* resources (uranium-235) is very similar to that with natural gas. In 2025 uranium of the first three categories is used, in 2050 — from the second to sixth category, in 2075 – from the third to the eight and in 2100 — the remains of the last two categories. Unlike fossil fuel, natural uranium of all eight cost categories is produced. Starting in 2050 waste uranium-238 will also be used (in breeders).

Price formation for uranium is much more complicated than for fossil fuel. The prices depend on stages of natural uranium enrichment, fuel assemblies production, spent fuel reprocessing, etc. Besides, in the second half of the century, production and use of nuclear fuel for fast reactors that will be used along with thermal reactors will affect price formation. In this connection the authors are not confident to make a forecast on uranium prices. We can point out only a general trend of their increase in the 21st century and this growth will apparently be less than that for fossil fuels.

Analysis of depletion periods for different cost categories of oil, natural gas and coal allowed one to make an approximate long-term forecast of the world (for natural gas — regional) prices for all these fuel kinds. Such a forecast is, naturally, possible only at several assumptions (initial prerequisites):

1. The forecast is developed as a range of price values.

2. The cost of the most expensive fuel categories used in the considered year is taken as a basis (the cost of the "closing" category). For coal the costs of its production in the exporting regions are taken into account.

3. "Weighted average" cost of the fuel produced can also be taken into account at a great variety of the categories used.

4. A definite excess of the fuel price over the cost of its production is supposed.

5. Fuel shortage is taken into account (for instance, oil shortage in 2100).

6. A fixed proportion is observed among the prices of oil, natural gas and coal that corresponds to the fuel quality (its consumer effect) and historically formed relations.

The forecasted prices based on these prerequisites are presented in Table 7.10.

Table 7.10. Long-term forecast of fuel prices, US$/tce

Fuel	Year				
	2000	2025	2050	2075	2100
Oil	130–150	160–180	190–220	240–280	270–320
Natural gas (EU)	80–90	80–100	100–130	150–190	210–260
Liquefied gas (JK)	100–110	100–120	120–150	170–210	230–280
Coal	40–50	50–60	60–80	70–100	80–120

It can be seen that an insignificant rise in prices is likely in the first quarter of the century; by the middle of the century the increase will be evident and in the second half of the century, as the cheap oil and gas resources are depleted, a sharp increase can be expected. Coal prices grow immensely though still remain 2–3 times lower than the prices of oil and gas (in energy equivalent).

7.3. Synthetic fuel production

The problem of synthetic fuel production was investigated in terms of several aspects:

1. Determination of *the appropriate time* for a transition to synthetic energy carriers that is caused by depletion of cheap conventional liquid and gaseous fuel resources.

2. Assessment of *the most attractive forms of synthetic fuel*, i.e. their ranking by economic efficiency, considering the whole process of their production and subsequent usage as well as the changing cost of primary energy resources they are produced from in the 21st century.

3. Determination of a mix of the most efficient methods of producing these priority forms of synthetic fuel, i.e. *ranking of their production methods.*

Consideration of these aspects for the eight Scenarios, for which the runs on the model GEM-10R were performed, would need much space and be redundant because of the coinciding analysis results for some Scenarios. Therefore, only two Scenarios (5 and 8) were selected. Scenario 5 was chosen in addition to our basic

Scenario 8 due to the fact that it is a most heavy Scenario in terms of external conditions and is characterised by a large-scale synthetic fuel production.

The synthetic fuels considered in GEM-10R are methanol, substitute natural gas (SNG), hydrogen and gasoline (representing in the aggregate all kinds of motor fuel), produced from coal or methanol.

Figure 7.3 presents five forms of secondary (converted) fuel and the methods of their production from primary energy and utilisation for final energy production that are foreseen in the model.

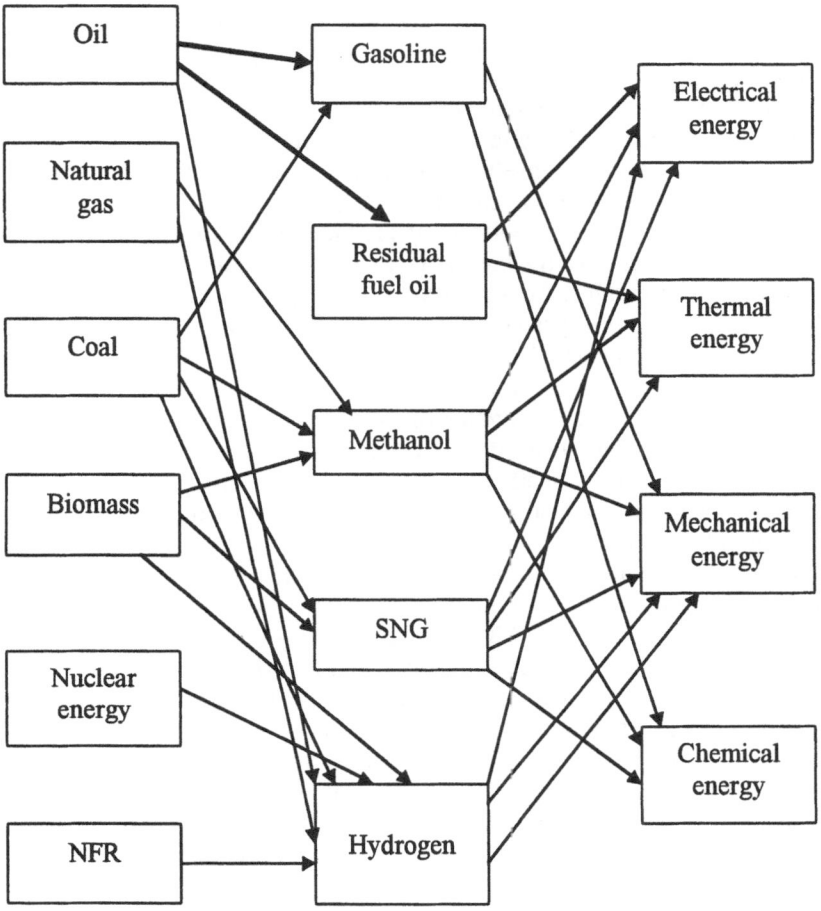

Figure 7.3. Methods (technologies) for production of secondary energy carriers and their subsequent utilisation.

For the purpose of simplification the figure does not show links which bear no relationship to the subject of this analysis, i.e. the technologies of direct use of primary energy for final energy production, use of electric energy as a secondary energy carrier (for example, for production of hydrogen, thermal and mechanical energy), use of methanol for gasoline production and some others.

Only two technologies (the thick lines) of those shown in the upper left-hand part of Figure 7.3 are conventional. The rest of them are intended for synfuel production. It is precisely the technologies which will be analysed and ranked based on the calculation results for Scenarios 5 and 8.

The amounts of primary energy used to produce the considered five forms of secondary fuel which are obtained from the runs for Scenarios 5 and 8 are given in Table 7.11. Analysis of this Table reveals the following trends.

The conventional conversion of oil to gasoline and residual fuel oil peaks in 2050 and practically stops by the end of the century due to depletion of cheap oil categories. Synthetic fuel production starts in 2050, except for nuclear-based hydrogen production in Scenario 5 (with rigid constraints on CO_2 emissions) and natural gas conversion to methanol in Scenario 8. For these conditions their production should commence as early as in 2025.

Gasoline (hydrocarbon motor fuel) produced from coal in large amounts in both Scenarios seems to be the most attractive synfuel. This shows that it is cheaper than conventional gasoline from the sixth cost category of oil, whose production is inefficient. In Scenario 8 gasoline from coal is the main synthetic fuel to be used even under the most favourable conditions envisaged in this Scenario.

Methanol, which is also produced in both Scenarios, ranks next to gasoline. GEM-10R employs several technologies for its production (see Figure 7.3). Based on the runs of the model their efficiency may be ranked in the following way: it is most profitable to produce methanol from natural gas, next from biomass and then from coal; methanol production from oil is inexpedient (it is more advisable to produce gasoline from oil).

Hydrogen (the third most attractive synthetic energy carrier) has to be used in very large quantities, at the rigid constraints on CO_2 emissions, for electricity generation as well (see Table 7.8). In small amounts it will be used with moderate constraints on CO_2 emissions (in Scenarios 6 and 7 by the end of the century).

In Scenario 5 (see Table 7.11) hydrogen is produced from a large number of energy resources. Its production from nuclear energy is most effective. This technology is followed by hydrogen production from oil (until it is depleted) and biomass and by the end of the century from coal and natural gas. The efficiency of these technologies differs by region. Hydrogen production from solar energy, energy from space and electricity, is the least advantageous primarily due to a high cost of these energy forms.

Substitute natural gas (methane) produced from coal or biomass is not used even in Scenario 5. Hence, it is the least effective synthetic fuel, particularly with constraints on CO_2 emissions (because of carbon presence). However, this conclusion should be treated as a general (global) trend. In the individual regions (or countries) having no natural gas but abundant cheap coal, production of substitute natural gas for heat supply may prove to be effective.

Table 7.11. Conversion of energy resources for fuel production, million TJ/year

Technology	Scenario 5 (HRM)				Scenario 8(LSM)			
	2025	2050	2075	2100	2025	2050	2075	2100
Oil to gasoline and residual fuel oil	163.5	206.9	172.7	0.8	153.4	185.9	140.9	37.0
Oil to hydrogen	0.0	23.6	0.0	0.0	0.0	0.0	0.0	0.0
Oil to methanol	0.0	0.0	0.0	0.0	0.0	0.0	0.0	0.0
Gas to methanol	0.0	7.0	45.7	58.2	1.3	22.4	12.7	9.6
Gas to hydrogen	0.0	0.0	0.0	96.7	0.0	0.0	0.0	0.0
Coal to gasoline	0.0	0.0	42.3	228.3	0.0	19.0	117.4	230.6
Coal to SNG	0.0	0.0	0.0	0.0	0.0	0.0	0.0	0.0
Coal to methanol	0.0	0.0	0.0	0.0	0.0	0.0	0.0	0.0
Coal to hydrogen	0.0	0.0	125.7	72.2	0.0	0.0	0.0	0.0
Uranium-235 to hydrogen	15.9	14.8	0.0	0.0	0.0	0.0	0.0	0.0
Uranium-238 to hydrogen	0.0	37.5	84.8	119.7	0.0	0.0	0.0	0.0
Biomass to methanol	0.0	0.0	0.0	0.0	0.0	0.0	0.0	9.5
Biomass to hydrogen	0.0	33.1	24.5	20.5	0.0	0.0	0.0	0.0
Biomass to SNG	0.0	0.0	0.0	0.0	0.0	0.0	0.0	0.0
Methanol to gasoline	0.0	0.0	0.0	0.0	0.0	0.0	0.0	0.0
LPS to hydrogen	0.0	0.0	0.0	0.0	0.0	0.0	0.0	0.0
Electricity to hydrogen	0.0	0.0	0.0	0.0	0.0	0.0	0.0	0.0
Solar energy to hydrogen	0.0	0.0	0.0	0.0	0.0	0.0	0.0	0.0
T o t a l	179.4	322.9	495.7	596.4	154.7	227.3	271.0	286.6

7.4. Use of energy carriers for final energy production

Technologies of final energy consumption were partially represented in the right-hand side of Figure 7.3 and besides they include technologies for direct use of primary energy resources and electricity to produce four kinds of final energy (electrical, thermal, mechanical and chemical) which were accepted in GEM-10R. Of concern are the possible changes in the mix and the efficiency of these technologies during the 21st century, considering the whole combination of conditions and factors which influence the energy structure in the regions and the world in general. The same two Scenarios (5 and 8), describing the unfavourable and favourable external conditions for energy development in the 21st century will be analysed.

Consumption of different primary energy resources and secondary energy carriers for *electricity production* for all eight Scenarios is shown in an aggregate way in Table 7.8. Its analysis reveals the following trends.

In Scenarios 1 and 2, without constraints on CO_2 emissions, electricity is produced basically from coal and nuclear energy (if there are no constraints on it).

Introduction of constraints on CO_2 emissions at high energy consumption (Scenarios 3–6) causes:

— a sharp reduction (in comparison to other Scenarios) in fossil fuel consumption at power plants, in particular coal; as was repeatedly noted, this fact complicates implementation of these Scenarios;

— the necessity to use hydrogen for peak power generation;

— the development of renewable energy sources on a very large (unreal) scale, if the constraints on nuclear energy development are also imposed (Scenarios 4 and 5);

— a many-fold growth of the total primary energy consumption for electric power generation. This is explained by increased production of electric power for heat generation, CO_2 removal, etc., on the one hand, and by lower efficiency of electric power production from nuclear energy and renewable energy sources, on the other hand.

In Scenarios 7 and 8 at low energy consumption and moderate (or even soft) constraints on CO_2 emissions and nuclear energy development, the structure of energy resources used for electricity production and its variation during the century become more realistic, especially in Scenario 8.

Note that in the optimal solutions obtained by runs on the model GEM-10R, natural gas is used at power plants in small volumes (particularly in 2025), as a rule, in lesser volumes than in 1990. This is due to the fact that natural gas as a high-grade fuel should be used by the least qualified consumers, such as heat consumers (including cooking). In the electric power industry natural gas should be used at peak power plants and cogeneration plants located in the cities (however, not in all regions).

Table 7.12 presents detailed calculation results on the application of different technologies of power production in Scenarios 5 and 8.

Table 7.12. Application of different technologies for electricity generation, million TJ/year

Technology	Scenario 5 (HRM)					Scenario 8 (LSM)			
	1990	2025	2050	2075	2100	2025	2050	2075	2100
Residual fuel oil (base*)	0.0	0.0	0.0	0.0	0.0	0.0	0.0	0.0	0.0
Residual fuel oil (peak**)	13.3	25.8	17.9	0.0	0.0	21.2	29.6	24.6	32.1
Gas (base)	4.2	1.4	0.0	0.0	0.0	0.0	0.0	0.0	0.0
Gas (peak)	10.0	1.3	17.1	0.0	0.0	1.0	1.6	9.1	6.1
Coal	50.5	11.7	0.0	0.0	0.0	104.6	119.6	112.7	82.1
Methanol (base)	0.0	0.0	0.0	0.0	0.0	0.0	0.0	0.0	0.0
Methanol (peak)	0.0	0.0	0.0	0.0	0.0	0.0	0.0	0.0	0.0
Uranium-235	23.2	43.6	67.8	37.7	21.0	43.4	80.2	46.9	28.3
Uranium-238	0.0	0.0	43.4	84.7	119.7	0.0	6.2	58.2	119.7
Biomass	0.8	19.6	0.0	0.0	0.0	0.0	0.0	0.0	0.0
Hydrogen (base)	0.0	0.0	0.0	0.0	0.0	0.0	0.0	0.0	0.0
Hydrogen (peak)	0.0	0.0	12.1	70.2	133.4	0.0	0.0	0.0	0.0
Existing HPPs	9.3	9.3	23.4	26.2	26.2	9.3	21.7	21.8	22.9
Hydro-1	0.0	11.9	0.0	0.0	0.0	11.9	0.0	0.0	0.0
Hydro-2	0.0	6.8	1.7	0.0	0.0	0.5	0.1	1.1	6.5
Hydro-3	0.0	2.2	1.1	0.0	0.0	0.0	0.0	0.0	0.0
Helio-1	0.0	23.2	24.1	105.5	127.9	0.0	0.0	0.0	0.0
Helio-2	0.0	0.0	2.7	0.0	0.0	0.0	0.0	0.0	0.0
Helio-3	0.0	0.0	0.0	0.0	0.0	0.0	0.0	0.0	0.0
Wind-1	0.1	7.8	9.0	9.0	9.0	0.3	9.0	9.0	9.0
Wind-2	0.0	50.8	57.6	62.1	62.0	0.0	0.0	0.0	8.8
Wind-3	0.0	17.2	19.0	42.0	42.0	0.0	0.0	0.0	0.1
Geo-1	0.0	6.7	11.6	14.1	14.0	0.0	0.0	0.0	0.0
Geo-2	0.0	6.0	10.4	18.1	0.0	0.0	0.0	0.0	0.0
Geo-3	0.0	0.0	0.0	0.0	0.0	0.0	0.0	0.0	0.0
LPS	0.0	0.0	24.9	67.4	265.4	0.0	0.0	0.0	0.0
Total	111.4	245.3	343.8	536.0	820.6	182.2	268.0	283.4	315.6

* Base load conditions of TPP.
** Peak load conditions.

Its analysis shows that:
— fuel oil, natural gas and hydrogen (in Scenario 5) are burnt only at peak power plants;

— nuclear energy is used roughly in the same volumes in both Scenarios within the imposed (moderate) constraints;

— water resources of the first two categories and the cheapest category of wind energy are the most effective renewable energy sources; they are used under all conditions;

— in Scenario 5 almost all renewable resources (excluding the most expensive categories of solar and geothermal energy) have to be brought into use;

— in the mid-century the second category of solar energy (terrestrial) and LPS are equally economic; by the end of the century, when the LPS cost is expected to fall considerably, the latter ranks below only the cheapest category of terrestrial solar energy;

— in Scenario 5 in the second half of the century the total consumption of primary energy for power production is two and more times higher than in Scenario 8, due to the reasons mentioned.

Figure 7.4 demonstrates dynamics of the installed capacities of power plants for Scenarios 5, and 8. In Scenario 5 the capacity of power plants in the end of the century is almost thrice as high as in Scenario 8. This increase is associated with several reasons: higher final demands for electric energy (at a high energy consumption level — compare Tables 3.18 and 3.21 in Section 3.4), large-scale use of electric energy for heat production (Table 7.13), the necessity to back up installations on renewable energy resources which are to be used on a large scale in Scenario 5 and others. Naturally it will be too difficult to provide such capacities of power plants as calculated for Scenario 5, which indicates once again desirability of reducing energy consumption and mitigating the constraints on CO_2 emissions.

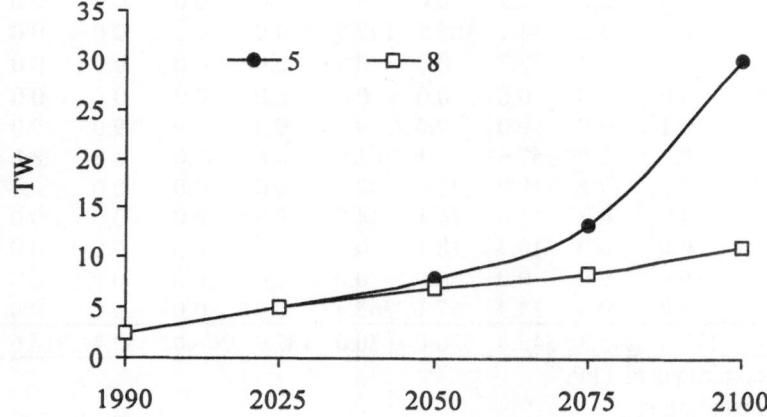

Figure 7.4. Installed capacity of power plants in the world as a whole in Scenarios 5 and 8.

Table 7.13 illustrates the use of different energy resources and energy carriers for *thermal energy production*. As distinct from the electrical energy only one generalised technology for heat production is foreseen for each energy carrier.

Table 7.13. Energy carriers consumption for heat production, million TJ/year

Energy carriers	Scenario 5 (HRM)					Scenario 8 (LSM)			
	1990	2025	2050	2075	2100	2025	2050	2075	2100
Gas	58.0	126.9	173.7	82.4	75.5	118.0	131.2	156.9	131.0
Coal	46.8	59.3	8.5	0.0	0.0	58.2	46.1	47.9	93.4
Uranium-235	0.0	0.4	2.7	4.8	6.0	0.0	0.0	0.0	0.0
Biomass	36.7	45.2	57.1	65.7	69.7	42.4	51.9	59.2	67.6
Residual fuel oil	31.2	24.0	30.4	25.6	0.1	22.5	25.2	10.2	4.8
Methanol	0.0	0.0	0.0	0.0	0.0	0.0	0.0	0.0	0.0
Hydrogen	0.0	0.0	33.9	153.4	15.8	0.0	0.0	0.0	0.0
Electrical energy	3.5	3.5	3.5	33.8	232.2	3.5	3.5	3.5	3.5
Total	176.2	259.3	309.8	365.7	399.3	244.6	257.9	277.7	300.3

The overall use of energy resources for heat production in Scenario 5 with a high energy consumption level is 20–30 percent larger than in Scenario 8. The individual energy carriers are characterised as follows:

— natural gas is used for heat supply on a large scale and a considerable fraction of its production (compare Tables 7.13 and 7.7 for Scenarios 5 and 8);

— in Scenario 5 coal is not used from the mid-21st century (because of constraints on CO_2 emissions);

— in the same Scenario heat supply is also based on nuclear energy, but in small amounts (in the regions of Europe and the former USSR);

— biomass and fuel oil are used in both Scenarios nearly in equal volumes (typical of the other Scenarios as well);

— in Scenario 5 hydrogen is used in large quantities in 2050 and 2075 and then in the end of the century is replaced by electrical energy produced from renewables.

On the whole the structure of energy carriers intended for heat production in Scenario 8 seems much more reasonable than in Scenario 5.

Consumption of different energy carriers for *mechanical energy production* is shown in Table 7.14. Here also one generalised technology is foreseen for each energy carrier (although the technologies have surely certain distinctions, for example for different types of transport: automobile, railway, sea, etc.).

Gasoline representing all types of motor fuel is used in both Scenarios to the greatest extent. Note that in the second half of the century it is produced basically from coal (see Table 7.11). Natural gas and methanol, which are practically not applied in the beginning of the century, also play a noticeable role. In Scenario 5 hydrogen is used in the second half of the century (about 10 percent of the total primary energy consumption).

Table 7.14. Energy carriers consumption for mechanical energy production, million TJ/year

Energy carriers	Scenario 5 (HRM)					Scenario 8 (SLM)			
	1990	2025	2050	2075	2100	2025	2050	2075	2100
Gasoline	62.1	91.4	108.1	104.3	96.4	85.9	111.1	139.9	130.2
Natural gas	0.0	3.3	11.7	19.2	30.1	7.2	12.6	19.2	30.1
Methanol	0.0	0.0	4.4	29.8	39.6	0.7	14.0	0.0	12.2
Hydrogen	0.0	0.0	4.8	15.9	24.9	0.0	0.0	0.0	0.0
Electric energy	0.1	0.1	5.3	8.6	13.2	0.1	0.8	0.1	2.6
Total	62.2	94.8	134.3	177.8	204.2	93.9	138.5	159.2	175.1

As regards *chemical energy production*, the use of three energy carriers: natural gas, gasoline (liquid hydrocarbons) and methanol is envisaged in the model GEM-10R for the chemical industry.

In all the Scenarios the consumption structure of energy carriers as feedstock for the chemical industry proved to be almost the same: gasoline is primarily used for these purposes till the mid-century and natural gas is used to meet increasing demands in the second half of the century. The gasoline consumption scale remains roughly at the level of 2050, though in the end of the century its larger portion is produced from coal. Table 7.15 presents consumption of energy carriers in the chemical industry for Scenarios 5 and 8.

Table 7.15. Energy carriers consumption for chemical energy production, million TJ/year

Energy carriers	Scenario 5 (HRM)					Scenario 8 (LSM)			
	1990	2025	2050	2075	2100	2025	2050	2075	2100
Gasoline	8.6	15.7	27.4	30.3	29.2	14.6	21.4	23.42	21.0
Natural gas	0.0	0.8	1.2	18.0	36.1	2.6	13.2	2.3	38.3
Total	8.6	16.5	28.6	48.3	65.3	17.2	34.6	45.7	59.3

Chapter 8

TENDENCIES IN ENERGY DEVELOPMENT OF WORLD REGIONS AND IN INTERREGIONAL TIES

8.1. Peculiarities in regional energy structures and interregional fuel exchange

The structure of energy resources consumption and conversion, their export and import and trends in individual world regions will be analysed for Scenario 8, which seems to be most realistic, as already noted. Table 8.1 presents consumption of energy carriers (both primary and converted) for all regions of the world that was obtained for this Scenario by applying the GEM-10R model. Liquid fuels include gasoline and fuel oil produced from oil (conventional), synthetic motor fuel from coal and methanol (crude oil is not used). Gaseous fuels comprise natural gas, its synthetic substitute (SNG) and hydrogen. As shown in Section 7.3 (see Table 7.11), hydrogen and SNG are not used in Scenario 8. Hence, in this case natural gas is the only gaseous fuel (besides, its small portion that is not considered in Table 8.1 is converted to methanol). For coal Table 8.1 presents amounts of coal directly used for production of final energy forms (electrical and thermal). A large fraction of coal converted to gasoline is not included in this table. Nuclear and renewable energy sources are given in volumes, totally consumed for electrical and thermal energy production (renewables include biomass).

Thus, liquid fuels in Table 8.1 are represented only by converted (secondary) energy carriers; gaseous ones include only natural gas (without its converted portion); coal is represented only by the portion of direct use for combustion and nuclear energy and renewables are primary energy resources. The total volumes of consumed liquid, gaseous fuels and coal (for the whole world) are lower than those of oil, natural gas and coal production (see Table 7.7, Section 7.1) by the value of losses during their transportation and conversion.

Analysis of Table 8.1 reveals that total consumption of energy carriers in industrially developed regions (NA, EU, JK and AZ) remains almost stable over the whole century, differing negligibly from consumption in 1990, decreases in SU and grows progressively in the other regions, particularly in Africa, China and South and Southeast Asia.

The use of liquid fuels in developed regions decreases, especially in the former USSR, grows significantly in developing regions by the mid-century and then stabilises. Consumption of natural gas grows in all regions (except for North America and the former USSR), particularly in AF, CH and SA. As for coal, the situation is more diversified: in some regions it is consumed rather steadily (NA, EU and AZ), in two regions its use falls (JK and SU), in the other regions it increases.

The use of nuclear energy and renewables increases in all regions. This is especially typical of nuclear energy in the regions of developing countries. Increase in the use of renewable energy sources is more modest, though rather high (two-threefold and almost fivefold in the former USSR).

Table 8.1. Consumption of energy resources and energy carriers, million TJ/year (Scenario 8 — LSM)

Year	Region										
	NA	EU	JK	AZ	SU	LA	ME	AF	CH	SA	World
					Liquid fuel						
1990	34.0	25.4	12.6	1.9	15.9	9.2	7.0	1.5	4.2	3.6	115.3
2025	27.4	21.2	5.6	0.8	5.5	7.9	5.6	11.5	38.3	20.9	144.7
2050	30.1	19.3	4.9	0.8	4.7	13.8	10.7	18.5	40.4	58.2	201.4
2075	22.2	19.7	4.6	0.8	3.4	13.5	9.5	21.6	39.2	63.7	198.2
2100	26.1	17.0	3.8	0.8	3.1	8.9	11.5	24.2	41.6	63.2	200.2
					Gaseous fuel						
1990	20.9	13.4	1.9	0.8	23.9	3.4	4.9	0.2	0.5	2.3	72.2
2025	22.5	18.9	6.4	0.9	24.7	6.3	5.6	12.1	8.6	14.0	126.1
2050	20.6	19.1	8.3	1.1	22.6	12.0	14.7	11.4	17.5	20.8	148.1
2075	29.2	21.4	9.0	1.0	23.7	15.9	19.4	15.8	36.8	26.3	198.5
2100	22.0	20.4	6.5	0.9	24.9	16.3	21.7	17.4	30.6	28.8	189.5
					Coal						
1990	21.3	20.5	4.1	1.7	15.3	0.9	0.3	2.8	24.2	6.1	97.2
2025	24.8	19.1	2.4	1.5	5.5	4.8	9.7	17.0	33.4	44.7	162.9
2050	23.1	23.0	0.0	0.2	3.8	1.5	11.1	23.2	40.7	39.2	165.8
2075	23.5	22.6	0.0	0.4	7.3	2.2	10.0	22.1	35.3	37.3	160.7
2100	13.9	16.8	2.2	1.6	3.2	6.0	8.5	29.6	47.1	46.8	175.7
					Nuclear energy						
1990	7.9	8.9	3.0	0.0	2.6	0.1	0.0	0.1	0.4	0.1	23.1
2025	8.7	9.1	8.1	0.0	4.1	0.5	0.2	0.4	6.9	5.4	43.4
2050	9.6	8.2	7.8	0.2	3.3	6.7	4.8	9.7	17.1	19.0	86.4
2075	8.0	8.5	7.4	0.4	4.2	8.6	7.9	14.2	21.0	25.0	105.2
2100	15.5	11.2	6.2	0.7	5.5	10.5	9.9	20.1	29.3	39.0	147.9
					Renewables						
1990	5.4	3.8	0.5	0.3	2.2	10.0	0.6	8.0	4.1	11.9	46.8
2025	5.6	4.4	0.9	0.6	3.4	15.0	1.3	10.8	7.9	14.5	64.4
2050	9.2	5.7	1.1	1.6	3.5	17.2	1.4	13.8	10.2	18.7	82.4
2075	9.4	6.0	1.1	1.6	6.2	18.2	1.3	13.8	10.2	22.8	90.6
2100	10.6	7.2	1.3	1.7	13.1	22.3	1.9	15.0	11.1	29.1	113.3
					Total						
1990	89.5	72.0	22.1	4.7	59.9	23.6	12.8	12.6	33.4	24.0	354.6
2025	89.0	72.7	23.4	3.8	43.2	34.5	28.5	51.8	95.1	99.5	541.5
2050	92.6	75.3	22.1	3.9	37.9	51.2	42.7	76.6	125.9	155.9	648.1
2075	92.3	78.2	22.1	4.2	44.8	58.4	48.1	87.5	142.5	175.1	753.2
2100	88.1	72.6	20.0	5.7	49.8	64.0	53.5	106.3	159.7	206.9	826.6

It should be noted that some "jumps" in figures of Table 8.1 for some regions and energy carriers are explained by the specific features of GEM-10R which were underlined in Chapter 2, in particular its static (or merely quasi-dynamic) character. In successive optimisation of each 25-year period the results obtained for individual regions do not always provide a smooth link with the neighbouring 25-year periods. This feature of GEM-10R is displayed to a greater extent with respect to the scales of energy carriers export and import (which will be considered below).

Table 8.2 presents scales of fuel conversion for Scenario 8. In this Scenario, fuels are converted to produce only liquid fuels. It is seen that in the first half of the century they are produced primarily from oil. In the second half, oil conversion diminishes due to depletion of its cheap resources, and synthetic gasoline becomes a major kind of fuel. Starting in 2025 natural gas is converted to methanol in the SU region, where natural gas resources are especially large. Besides, in 2100 methanol is produced on a small scale from biomass (see Table 7.11, Section 7.3) chiefly in Latin America, which is not shown in Table 8.2.

The runs on GEM-10R show that it is more economical to convert oil in the regions of its production and export oil products (rather than crude oil). Therefore, oil is refined in all regions where it is produced. The Middle East, the former USSR and Latin America refine it on the greatest scale.

Coal conversion to gasoline starts in 2050 in North America and Europe, followed by China, the former USSR and Australia. Coal conversion reaches the largest scales by the end of the century. This is the most economical way of meeting demands for liquid fuel after cheap oil reserves have been depleted.

As noted, conversion of natural gas to methanol starts in the year 2025 in the former USSR and a relatively small-scaled production of methanol from biomass emerges in the very end of the century. Methanol is used basically for meeting domestic demands for liquid fuel, as far as its transport is less efficient than that of gasoline.

Table 8.3 indicates regional export and import of fossil fuels for the same Scenario. In its analysis, account should be taken of the following important circumstance. In the model runs there were no constraints on the volumes of fuel export and import. The exception is the natural gas export from the former USSR and its import to China and South and Southeast Asia that were determined by the authors by expert judgement on the base of actual possibilities for such export. Fuel production volumes by region (except for setting the volumes of resources by cost category) were also not limited. Therefore, the model freely selected the volumes of fuel production and export/import based on the minimum cost criterion for the whole world. The results obtained were optimal for the world energy system, but for individual regions they often appeared to be unrealistic. Factually each region is characterised by numerous specific financial-economic, environmental and socio-political conditions and factors that will inevitably limit fuel export and import and its production.

This fact is a serious disadvantage in the computations performed and the results obtained (figures in Table 8.3) should be treated in a special way. They must be considered as a reflection of general optimal trends which will manifest themselves in the future. They must not be interpreted as a specific forecast, they rather can be applied for qualitative analysis, i.e. identification of tendencies in the interregional fuel exchange. The analysis of this type is performed below for each of the ten regions.

Table 8.2. Fuel conversion, million TJ/year (Scenario 8 — LSM)

Year	Region										
	NA	EU	JK	AZ	SU	LA	ME	AF	CH	SA	World
Oil to oil products											
1990	34.0	26.0	12.6	1.6	15.9	9.2	7.0	1.6	4.2	4.6	116.5
2025	9.9	5.2	0.0	0.7	26.8	22.6	71.8	7.6	6.5	2.2	153.4
2050	20.8	6.4	0.0	0.9	35.9	26.5	76.8	9.2	9.6	0.0	185.9
2075	0.0	0.0	0.0	0.0	23.0	23.5	90.3	0.0	4.1	0.0	140.9
2100	7.1	0.0	0.0	0.0	5.1	4.0	20.8	0.0	0.0	0.0	37.0
Gas to methanol											
1990	0.0	0.0	0.0	0.0	0.0	0.0	0.0	0.0	0.0	0.0	0.0
2025	0.0	0.0	0.0	0.0	1.3	0.0	0.0	0.0	0.0	0.0	1.3
2050	0.0	0.0	0.0	0.0	22.4	0.0	0.0	0.0	0.0	0.0	22.4
2075	0.0	0.0	0.0	0.0	12.7	0.0	0.0	0.0	0.0	0.0	12.7
2100	0.0	0.0	0.0	0.0	9.6	0.0	0.0	0.0	0.0	0.0	9.6
Coal to gasoline											
1990	0.0	0.0	0.0	0.0	0.0	0.0	0.0	0.0	0.0	0.0	0.0
2025	0.0	0.0	0.0	0.0	0.0	0.0	0.0	0.0	0.0	0.0	0.0
2050	15.4	3.6	0.0	0.0	0.0	0.0	0.0	0.0	0.0	0.0	19.0
2075	40.7	10.1	0.0	1.0	9.2	0.0	0.0	0.0	56.4	0.0	117.4
2100	24.3	31.6	0.0	64.0	47.1	0.5	0.0	0.0	63.0	0.0	230.6
Total											
1990	34.0	26.0	12.6	1.6	15.9	9.2	7.0	1.6	4.2	4.6	116.5
2025	11.2	5.1	0.0	0.7	26.8	22.6	71.8	7.6	8.8	2.2	154.7
2050	36.2	10.0	0.0	0.9	58.3	26.5	76.8	9.1	9.6	0.0	227.3
2075	40.7	10.1	0.0	1.0	44.9	23.5	90.3	0.0	60.5	0.0	271.0
2100	31.4	31.6	0.0	64.0	61.8	4.5	20.7	0.0	63.0	0.0	277.1

First of all, the fuel exporting and importing regions can be determined sufficiently well from the total volumes of export and import. As now, in the 21st century the Middle East, the former USSR, Latin America and Australia will remain chief fuel exporters. North America, which can shift from a net importer to a sufficiently large fuel exporter, occupies an intermediate position. The rest of the regions will import fuel. The SA region may import particularly large volumes.

Let us consider now basic trends in energy development for individual regions from the analysis of Tables 8.1-8.3.

Consumption structure of energy carriers in *North America* is rather stable over the whole century (Figure 8.1). Only the fraction of coal markedly decreases and that of nuclear energy and renewables, on the contrary, increases in the end of the century. Section 8.2 deals with utilisation of renewables in greater detail.

Table 8.3. Export (with sign "–") and import of energy carriers, million TJ/year (Scenario 8 – LSM)

Year	Region										
	NA	EU	JK	AZ	SU	LA	ME	AF	CH	SA	World
Liquid fuel											
1990	14.4	19.4	13.9	0.4	−4.9	−6.6	−38.0	−2.6	0.6	1.7	52.1
2025	14.8	13.5	5.9	0.1	−17.0	−13.0	−56.3	4.0	31.7	16.3	86.3
2050	0.0	10.7	4.1	−0.1	−43.0	−10.4	−57.9	10.2	29.7	56.8	111.5
2075	0.0	15.0	4.8	0.3	−37.4	−5.7	−67.2	23.3	0.0	67.2	110.3
2100	0.0	−1.5	3.9	−42.1	−36.7	3.5	−19.6	24.0	2.8	65.8	99.9
Gaseous fuel											
1990	0.6	4.0	2.2	−0.2	−3.5	−0.4	−2.5	0.0	0.1	−2.8	9.3
2025	−10.1	6.5	6.8	−0.6	−5.0	−2.7	−9.2	11.5	4.0	−1.2	28.8
2050	3.6	4.6	9.7	−1.4	−10.0	−6.2	−28.9	12.5	8.0	8.0	46.5
2075	0.0	21.9	9.8	0.0	−15.0	0.0	−48.3	7.7	12.0	12.0	63.3
2100	0.0	0.0	7.0	−7.9	−20.0	0.0	−18.3	0.0	20.0	19.2	46.2
Coal											
1990	−3.0	3.2	3.8	−3.3	−0.6	−0.4	0.1	−0.7	−0.1	0.6	8.1
2025	−25.0	0.0	2.4	−7.9	−22.4	0.0	9.7	8.8	0.0	34.5	55.3
2050	0.0	0.0	0.0	0.0	−11.1	0.0	11.1	0.0	0.0	0.0	11.1
2075	0.0	0.0	0.0	0.0	−48.9	0.0	10.0	0.0	0.0	38.9	48.9
2100	−27.4	0.0	2.2	−8.4	0.0	0.0	8.5	0.0	0.0	25.2	35.8
Total											
1990	12.1	26.5	19.9	−3.1	−9.0	−7.3	−40.4	−3.3	0.6	−0.5	63.6
2025	−20.3	20.0	15.1	−8.4	−44.4	−15.7	−55.8	24.3	35.7	49.5	144.7
2050	3.6	15.3	13.7	−1.5	−64.1	−16.6	−75.8	22.7	37.7	64.8	157.9
2075	0.0	36.8	14.6	0.3	−10.3	−5.7	105.5	31.0	12.0	118.1	212.5
2100	−27.4	−1.5	13.1	−58.4	−56.7	3.5	−29.4	24.0	22.8	110.2	173.4

In 2050 the region starts production of synthetic gasoline from coal (see Table 8.2), which leads to termination of liquid fuel import (see Table 8.3). Besides, the region exports coal and at some time intervals natural gas (the figures for export and import are unstable because of the reasons mentioned above). On the whole, the North America region is well provided with domestic energy resources; it may be counted as self-balancing and, depending on the situation that develops, it can be both fuel exporter and importer owing to stabilisation of the total consumption of energy resources from the very beginning of the century.

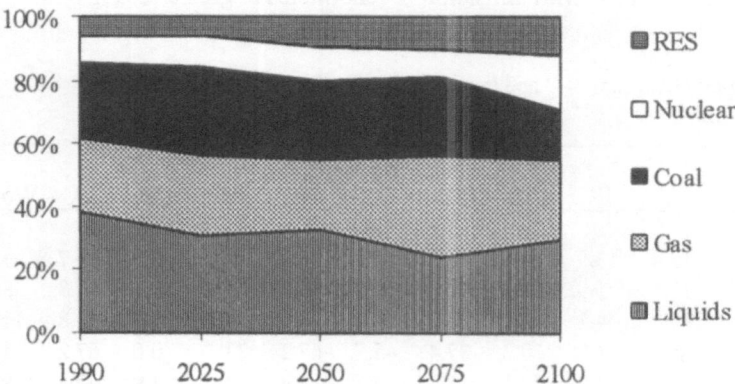

Figure 8.1. Structure of energy resources and energy carriers consumed in North America (Scenario 8 – LSM).

The structure of energy carriers used in the *Europe* region (Figure 8.2) is characterised by a fall in the fraction of liquid fuels and rise of the natural gas fraction. Utilisation of coal and nuclear energy, which are approximately of the same economic efficiency, can vary, but their total fraction remains constant. The fraction of renewables grows negligibly by the end of the century.

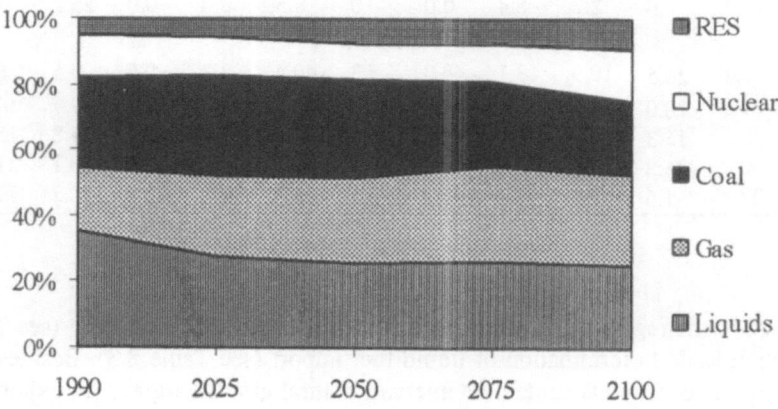

Figure 8.2. Structure of energy resources and energy carriers consumed in Europe (Scenario 8 — LSM).

As America, Europe begins production of motor fuel from coal in 2050, reducing requirement of its import. However, during almost the whole 21st century the region remains a large importer of liquid fuel and natural gas (despite its rather large production scales in the region). Only in the very end of the century the situation at the world fuel markets proves to be favourable, on the one hand, for production of natural gas of cost category 6, whose reserves in Europe are rather large and for conversion of domestic coal to liquid fuel, on the other hand. As a result it becomes possible to terminate gas and liquid fuel import and even export the latter in small amounts.

The *Japan and Republic of Korea* region is characterised by some reduction in total consumption of energy resources by the end of the century. The consumption structure changes to a great extent (Figure 8.3): the liquid fuel fraction falls considerably and coal is practically not used, however, the fractions of natural gas and nuclear energy grow to the same extent. The fraction of renewable energy sources increases, but does not exceed 5–7%.

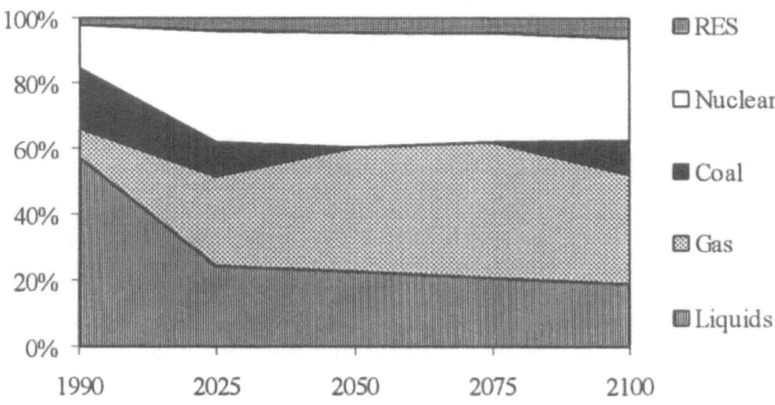

Figure 8.3. Structure of energy resources and energy carriers consumed in Japan and Korea (Scenario 8 – LSM).

As was noted above, in the optimal solution obtained on GEM-10R it is economically inefficient to export crude oil. Its refining in situ and export of oil products (fuel oil and gasoline) appear more advisable. Therefore, in the JK region there is no oil refining in the 21st century (see Table 8.2), though in 1990 it was refined and is refined now on a large scale there. Crude oil export and import and its refining in importing countries will apparently remain for some time, at least in the first half of the century and therefore corresponding adjustments must be made in Table 8.2. However, in the authors' opinion this is not so important. Significant is the fact that the JK region remains a large importer of all kinds of fossil fuel during the whole century, though the import fraction in total consumption of energy carriers (Tables 8.1 and 8.3) diminishes by the mid-century due to growing fractions of nuclear energy and renewables.

The *Australia and New Zealand* region, as the previous two developed regions, shows a decreasing tendency in consumption of liquid fuel and coal and an increasing one in utilisation of natural gas (Figure 8.4, see Table 8.1). The runs on GEM-10R reveal the expediency of an essential increase in the fraction of renewables (from 6–7% in 1990 to 30–40% in the second half of the century) and of using nuclear energy from the mid-century in the region. Now it is difficult to judge to what extent such energy development in the region is probable.

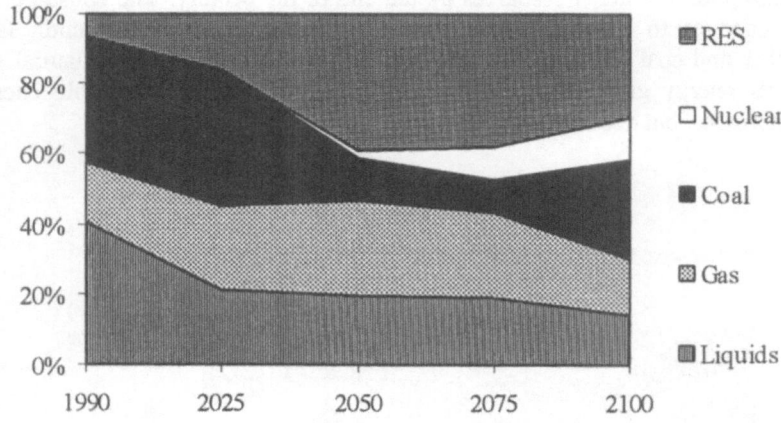

Figure 8.4. Structure of energy resources and energy carriers consumed in Australia and New Zealand (Scenario 8 — LSM).

The AZ region is a fuel exporting region. It exports basically coal and natural gas. The calculations, however, show that in the end of the century large-scale conversion of coal to liquid fuel and export of the latter become advisable here. Very high scales of export obtained in the model runs may cast some doubt, but this tendency of using coal in the late 21st century should be taken as economically valid.

The *former USSR* region demonstrates rather interesting and basically unexpected trends (Figure 8.5). First of all, in contrast to all other regions the total consumption of energy resources falls considerably here as compared with 1990 (see Table 8.1). This is due to the forecasted analogous decrease in energy consumption in the region (see Section 3.4). It presumes that the region will never reach the 1990 energy consumption volumes because of inevitable decrease in GDP-energy ratio for many reasons.

Consumption of liquid fuel falls especially drastically. The natural gas fraction grows with stable absolute amounts of its use. The fractions of nuclear energy and renewables also increase.

The results of optimisation calculations show large-scale conversion of oil and in the second half of the century natural gas and coal to liquid energy carriers in the region (see Table 8.2). Almost all these fuel kinds are exported. At the same time large volumes of natural gas and coal are also supplied for export. As a result the SU region may become a largest fuel exporter (along with the Middle East). However, such scaled fuel conversion and export in the region are hardly possible because of a combination of economic, environmental and socio-political factors underlined above.

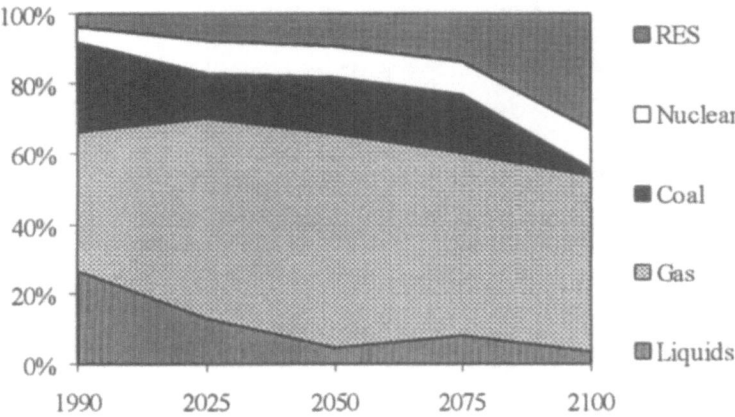

Figure 8.5. Structure of energy resources and energy carriers consumed in the former USSR (Scenario 8 — LSM).

Consumption structure of energy resources in the *Latin America* region is characterised by the following specific features (Figure 8.6):
   – the liquid fuel fraction falls, particularly by the end of the century;
   – the fractions of natural gas and nuclear energy grow;
   – coal consumption is unstable, but low and does not exceed, as a rule, 10%;
   – the fraction of renewable energy sources is high enough and stable, comprising some 40% in the first half of the century with its decrease to 30% in the end of the century.

The region exports oil products and natural gas in volume during most of the century. And only in the end of the century after depletion of cheap oil and gas resources it turns into an importer. In parallel the use of coal and nuclear energy increases, conversion of coal to gasoline and, as mentioned, biomass to methanol begins (Table 8.2).

On the whole, as distinct from the previously considered regions, consumption of energy resources in Latin America rises threefold by the end of the century in comparison to the year 1990 (see Table 8.1). In the other developing regions this growth, meanwhile, will be even greater. Note that for the Latin America region, economic assistance by the developed countries was not envisaged. Therefore, its energy (and economy) development in accordance with Scenario 8 is implemented, so to say, by its own means.

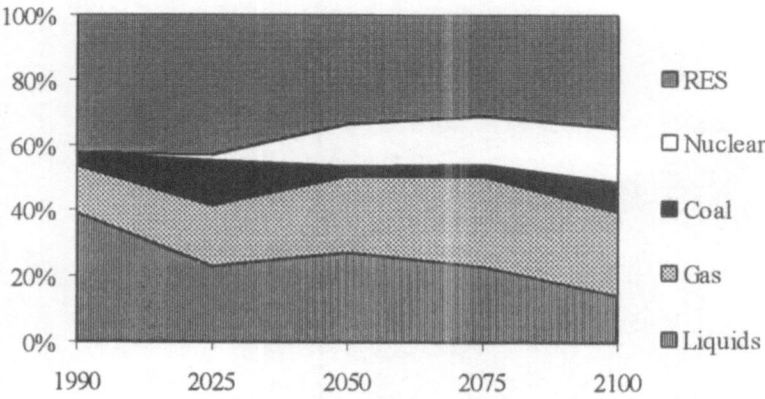

Figure 8.6. Structure of energy resources and energy carriers consumed in Latin America (Scenario 8 — LSM).

The *Middle East* region remains a main exporter of oil products over the whole 21st century and becomes a largest natural gas exporter. The structure of domestic consumption of energy resources is characterised by the decreasing share of liquid fuel and the increasing shares of natural gas and coal (Figure 8.7). The use of coal (though it is fully imported) at thermal power plants is economically expedient to raise the volumes of oil and natural gas export. The same reason underlies utilisation of nuclear energy (especially in the countries without oil and gas resources).

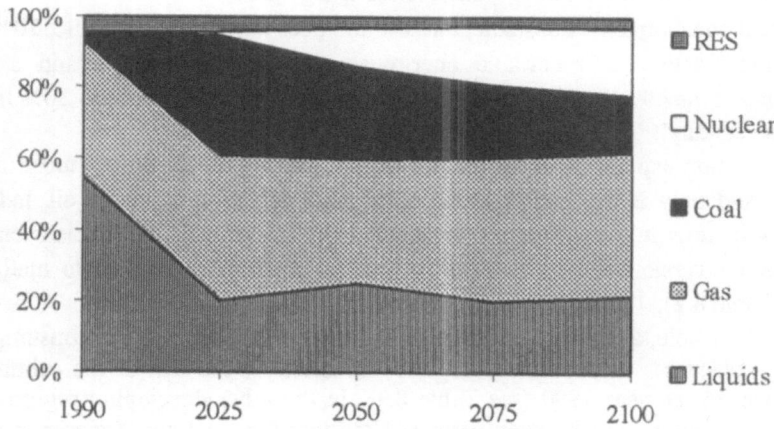

Figure 8.7. Structure of energy resources and energy carriers consumed in Middle East (Scenario 8 — LSM).

In this Scenario the use of renewables in the Middle East is negligible. However, in the Scenarios with rigid constraints on CO_2 emissions the terrestrial solar energy, whose resources here are large enough, should develop on a large scale.

Total energy consumption in the region grows by the end of the century more than fourfold. And the region requires some (not very essential) economic assistance (see Section 3.4).

The *Africa* region is economically least developed with the highest rates of population growth forecasts. As shown in Section 3.4, it will not be able to shift to a state sufficient for sustainable development in the early 21st century without substantial economic aid. Forecast of energy consumption applied in Scenario 8 is made on the assumption that such aid will be rendered. Consumption of energy resources in this case will grow by the end of the century more than eightfold (see Table 8.1). Such a growth leads to a multiple increase in the use of all kinds of energy carriers.

Increase in the fractions of natural gas and nuclear energy is particularly sharp (Figure 8.8). The fractions of coal and liquid fuel grow to a lesser extent. This occurs due to a sharp reduction in the share of renewables that was above 60% in 1990. Basically they are represented by non-commercial biomass (above 90%) that is used for heat production (food cooking).

Such consumption scales in Africa can be provided by import of large volumes of oil products, natural gas and coal (Table 8.3). From a small exporter (in 1990) it will turn into a large fuel importer during the whole current century. It should be underlined once again that in GEM-10R there were no constraints on fuel export and import, therefore resources for realisation of such imports in terms of financial, economic and other conditions have to be verified further on the basis of additional studies and analysis.

The *China* region will become a largest energy consumer (and producer) in the 21st century. Total consumption of resources will grow almost fivefold by the end of the century. Economic assistance to the region by the developed countries was not envisaged.

Despite the tremendous increase in energy consumption, China covers it mainly by domestic resources. Oil products imported on a large scale in the first half of the century and natural gas imported in lower volumes are an exception. In the second half of the century demands for liquid fuel are met by the scaled conversion of coal to gasoline.

Consumption structure of energy carriers is characterised by sharp increases in the fractions of liquid fuel and natural gas in the first half of the century and the fraction of nuclear energy in the second half (Figure 8.9). The coal fraction that made up above 70% in the year 1990 falls to 30% by the end of the century. The share of renewables also drops from 12 to 7%.

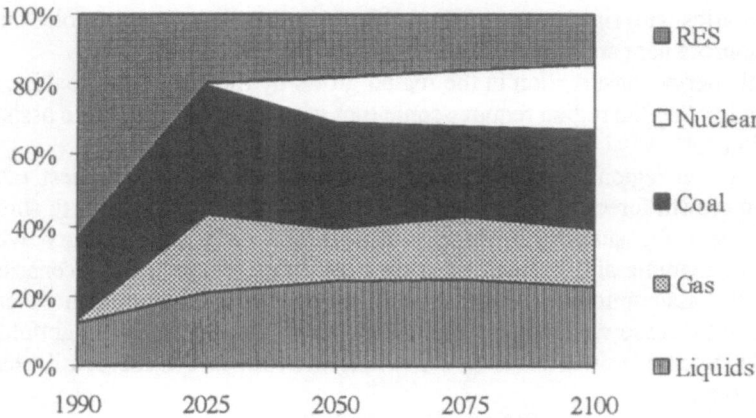

Figure 8.8. Structure of energy resources and energy carriers consumed in Africa (Scenario 8 — LSM).

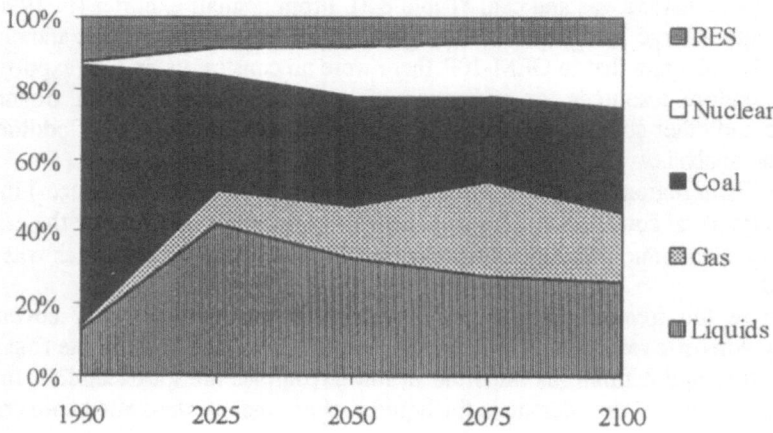

Figure 8.9. Structure of energy resources and energy carriers consumed in China (Scenario 8 — LSM).

Data for the *South and Southeast Asia* (SA) region are most impressive. This region has the largest population even now and will keep the first place over the whole 21st century. Sustainable development of the region is possible only with substantial economic aid. The energy consumption forecast made in terms of such an aid supposes increase in energy consumption by the end of the century more than fivefold in comparison with 1990.

In contrast to China this region is poorly provided with domestic energy resources and hence, must import about half of the energy resources consumed. Total fuel import to the region accounts for from 30 to 60% of its world export at different periods of the century. Here, as in the Africa region, the question arises of the possibility of importing such fuel volumes.

Consumption structure of energy resources and energy carriers for the SA region is characterised in the 21st century by the following trends (Figure 8.10). The fraction of renewables that reached almost 50% in 1990 falls in the end of the century to 12%. The fraction of remaining energy carriers increases: primarily liquid fuel, natural gas and coal in the first half of the century and nuclear energy in the second half.

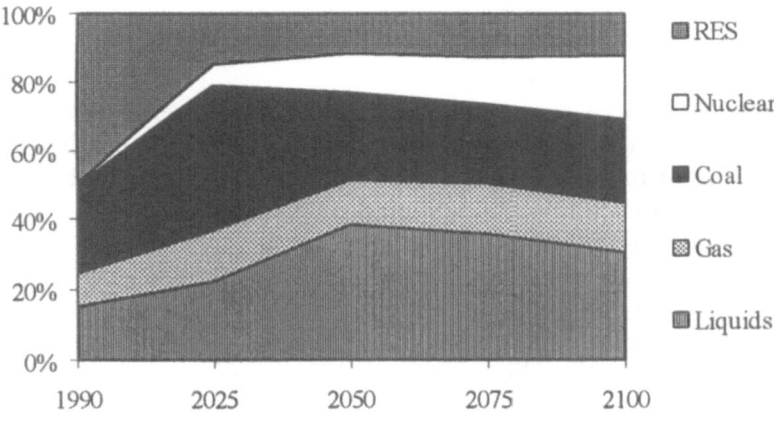

Figure 8.10. Structure of energy resources and energy carriers consumed in South and Southeast Asia (Scenario 8 — LSM).

8.2. Regional peculiarities in using energy carriers and technologies for electrical, thermal and mechanical energy production

Specific features of the regions were analysed for the same Scenarios 5 and 8 that were considered for the whole world in Section 7.4. As noted there, they cover virtually the whole range of favourable and unfavourable future conditions of energy development.

Let us consider at first energy carriers and technologies for *electrical energy* production. Section 7.4 described global trends to using fossil fuel, nuclear energy and the cheapest forms of renewable energy sources (hydropower and cheap wind resources) at power plants under favourable conditions (Scenario 8) and factually complete replacement of fossil fuel by renewable sources and hydrogen under unfavourable conditions. Differences among the regions lie primarily in the use of fuel oil or natural gas (at peak power plants) and renewables.

Tables 8.4a and 8.4b present electricity production structure in the regions by form of energy carriers (energy resources) for Scenarios 5 and 8. To reduce the size of tables these data are given only for 2050 and 2100 and only for the energy carriers that are used in the considered years and Scenarios. Absence of any energy carrier implies that it is used in none of the regions.

In favourable conditions (Table 8.4b) the peak power plants (apart from hydropower plants) burn fuel oil in the majority of the regions. Only in the former USSR region and also in Latin America in 2100 is use made of natural gas (Table 8.4b). With more rigid constraints on CO_2 emissions, fuel oil is replaced by the mid-century by natural gas and hydrogen (excluding China and South and Southeast Asia) and only by hydrogen in the end of the century in all regions. Its use commences at first (in 2050) in the regions of developed countries. Hence, fuel oil is the most economical fuel for peak power plants, then follows natural gas and at last, only in extreme cases, hydrogen.

Under favourable conditions coal as the cheapest resource (along with nuclear energy) is used for base power production in all regions, except for Japan and Republic of Korea (in the end of the century it is no longer used in the former USSR and Latin America as well) (see Table 8.4b). Under unfavourable conditions it is utilised in none of the regions in 2050.

Starting in 2050 nuclear energy is utilised in all regions in accordance with the established (moderate) constraints (as an upper limit) in any conditions. It must be used most intensively in China and South and Southeast Asia and in the end of the century in Africa, i.e. in the regions with insufficient fuel resources.

Utilisation of renewable energy sources is characterised by the following specific features:

– under favourable conditions (see Table 8.4b) cheap hydro and wind energy resources are in common use, especially in Latin America, the former USSR, North America, South and Southeast Asia, China and Europe;

– terrestrial solar energy is developed in unfavourable conditions (see Table 8.4a) in the Middle East and Africa (on a very large scale) and also in North and Latin America (on a small scale in China and South and Southeast Asia in 2050 as well);

— energy of space systems should be utilised in the same conditions at first in Europe, China, South and Southeast Asia (in 2050) and then in Japan (and Republic of Korea) and also in Africa; the required (and economically expedient for such conditions) scales of its use are much higher than the scales of terrestrial solar energy;

– use of wind energy increases considerably, if more rigid constraints are imposed on CO_2 emissions, particularly in North America, the former USSR and Latin America;

– under unfavourable conditions geothermal energy is used in all regions, but on a relatively modest scale.

Table 8.4a. Use of energy resources and energy carriers for electricity production, millionTJ/year (Scenario 5 — HRM)

Energy carrier	Region										
	NA	EU	JK	AZ	SU	LA	ME	AF	CH	SA	World
					1990						
Fuel oil	1.4	1.5	2.9	0.4	4.3	0.7	1.5	0.2	0.4	0.1	13.3
Natural gas	2.8	2.9	1.6	0.2	3.6	0.3	1.4	0.1	0.0	1.2	14.2
Coal	20.9	10.2	0.4	1.2	6.5	0.5	0.0	1.3	6.3	3.2	50.5
Biomass	0.0	0.0	0.0	0.0	0.0	0.3	0.0	0.2	0.0	0.3	0.8
Nuclear energy	7.9	8.9	3.0	0.0	2.6	0.1	0.0	0.1	0.4	0.1	23.2
Hydropower	2.4	2.1	0.4	0.2	1.0	1.7	0.1	0.2	0.7	0.5	9.3
Solar energy	0.0	0.0	0.0	0.0	0.0	0.0	0.0	0.0	0.0	0.0	0.0
Wind energy	0.1	0.1	0.0	0.0	0.0	0.0	0.0	0.0	0.0	0.0	0.1
Geothermal energy	0.0	0.0	0.0	0.0	0.0	0.0	0.0	0.0	0.0	0.0	0.0
LPS	0.0	0.0	0.0	0.0	0.0	0.0	0.0	0.0	0.0	0.0	0.0
Total	35.5	25.7	8.2	2.0	18.1	3.6	3.0	2.1	7.9	5.4	111.4
					2050						
Fuel oil	0.0	0.0	0.0	0.0	0.0	0.0	0.0	0.0	7.8	10.1	17.9
Natural gas	5.3	3.4	0.1	0.3	3.0	0.0	4.2	0.9	0.0	0.0	17.1
Hydrogen	3.6	3.5	1.5	0.1	0.0	3.5	0.0	0.0	0.0	0.0	12.1
Nuclear energy	15.6	15.2	8.0	0.3	4.8	4.2	4.8	3.5	24.0	31.0	111.3
Hydropower	2.7	2.4	0.4	0.2	5.2	8.7	0.3	1.3	2.6	2.5	26.2
Solar energy	0.0	0.0	0.0	0.0	0.0	0.0	24.1	0.0	0.3	2.4	26.8
Wind energy	31.7	7.0	0.8	1.5	22.0	7.7	0.7	1.4	6.6	6.2	85.6
Geothermal energy	2.5	2.0	2.1	0.4	0.9	0.0	0.3	2.0	1.8	10.0	22.0
LPS	0.0	5.9	0.0	0.0	0.0	0.0	0.0	0.0	8.5	10.5	24.9
Total	61.3	39.4	12.9	2.8	35.8	24.0	34.4	9.1	51.5	72.6	343.8
					2100						
Hydrogen	12.2	11.4	3.2	0.8	3.9	6.2	14.5	12.1	24.9	44.3	133.4
Nuclear energy	12.1	10.8	5.3	0.6	5.3	10.5	7.7	20.1	29.3	39.0	140.6
Hydropower	2.7	2.4	0.4	0.2	6.0	8.0	0.3	1.2	2.6	2.5	26.2
Solar energy	8.0	0.0	0.0	0.0	0.0	8.0	72.0	40.0	0.0	0.0	127.9
Wind energy	50.2	7.0	0.8	4.9	22.0	11.2	0.7	3.4	6.6	6.2	113.0
Geothermal energy	2.5	1.5	0.7	0.0	0.9	2.8	0.3	2.0	0.8	2.5	14.0
LPS	0.0	28.8	7.3	0.0	0.0	0.0	0.0	7.5	76.3	145.5	265.4
Total	87.6	61.9	17.7	6.5	38.1	46.7	95.4	86.2	140.5	240.0	820.6

It should be noted that energy from space is utilised either in regions with poor resources of terrestrial solar energy (Europe) or in regions that are characterised by long rainy seasons, during which the use of terrestrial solar energy for electricity production is obviously inefficient (the JK, CH and SA regions).

Table 8.4b. Use of energy resources and energy carriers for electricity production, million TJ/year (Scenario 8 — LSM)

Energy carrier	Region										
	NA	EU	JK	AZ	SU	LA	ME	AF	CH	SA	World
					2050						
Fuel oil	4.9	4.0	1.0	0.2	0.0	2.4	1.7	3.4	4.7	7.4	29.6
Natural gas	0.0	0.0	0.0	0.0	1.6	0.0	0.0	0.0	0.0	0.0	1.6
Coal	17.6	14.5	0.0	0.0	3.8	1.5	11.1	17.1	23.0	31.1	119.6
Nuclear en.	9.6	8.2	7.8	0.1	3.3	6.7	4.8	9.7	17.1	19.0	86.3
Hydropower	2.7	2.4	0.4	0.2	3.0	6.0	0.4	1.2	2.6	2.5	21.8
Wind energy	3.0	1.0	0.2	1.0	0.5	1.2	0.1	0.2	0.6	1.2	9.0
Total	37.7	30.0	9.5	2.0	11.8	17.8	18.0	31.5	48.0	61.2	268.0
					2100						
Fuel oil	4.7	3.5	1.0	0.3	0.0	0.6	2.2	4.4	6.3	9.2	32.1
Natural gas	0.0	0.0	0.0	0.0	2.8	3.3	0.0	0.0	0.0	0.0	6.1
Coal	8.4	8.3	0.0	1.5	0.0	0.0	8.5	13.8	19.9	21.8	82.1
Nuclear en.	15.5	11.2	6.2	0.7	5.5	10.5	9.9	20.1	29.3	39.0	147.9
Hydropower	3.7	3.4	0.6	0.3	4.2	9.8	0.3	1.2	3.6	2.5	29.4
Wind energy	3.0	1.0	0.2	1.0	4.2	1.2	0.6	1.4	0.6	4.7	17.9
Total	35.3	27.3	8.0	3.7	16.7	25.5	21.4	40.9	59.7	77.1	315.6

Tables 8.5a and 8.5b illustrate the use of energy carriers for *thermal energy* production in the regions. As the previous tables, they present data for 2050 and 2100 and the energy carriers that are applied in the corresponding Scenario.

Natural gas makes the greatest contribution to heat production and is utilised in all regions. Under favourable conditions its portion in the total consumption for heat supply dominates over the entire century (see figures for gas in Tables 8.1 and 8.5b). Even under unfavourable conditions (see Table 8.5a) natural gas meets more than 50% of thermal energy demand in 2050 and only in the end of the century is it substituted by electric power. Its highest specific weight in heat production is observed in the ME, SU and JK regions (up to 90% in Scenario 8).

Under favourable conditions coal is consumed on a considerable scale for heat production in North America, Europe, Africa, China, South and Southeast Asia. In the other regions its consumption for these purposes is negligible. Under unfavourable conditions coal is not used starting in 2050 universally (except for small volumes in China).

Table 8.5a. Use of energy resources and energy carriers for heat production, million TJ/year (Scenario 5 — HRM)

Year	NA	EU	JK	AZ	SU	LA	ME	AF	CH	SA	World
						Region					
						Natural gas					
1990	18.1	10.5	0.3	0.6	20.3	3.1	3.5	0.1	0.5	1.1	58.0
2050	29.9	29.3	8.8	1.2	17.3	16.5	20.1	0.0	18.5	32.2	173.7
2100	14.6	0.0	0.0	0.0	0.0	22.9	11.6	12.4	13.9	0.0	75.5
						Coal					
1990	0.4	10.3	3.8	0.6	8.8	0.4	0.3	1.5	17.8	1.1	46.8
2050	0.0	0.0	0.0	0.0	0.0	0.0	0.0	0.0	8.5	32.2	8.5
2100	0.0	0.0	0.0	0.0	0.0	0.0	0.0	0.0	0.0	0.0	0.0
						Nuclear energy					
1990	0.0	0.0	0.0	0.0	0.0	0.0	0.0	0.0	0.0	0.0	0.0
2050	0.0	1.2	0.0	0.0	1,5	0.0	0.0	0.0	0.0	0.0	2.7
2100	0.0	2.8	0.0	0.0	3.2	0.0	0.0	0.0	0.0	0.0	6.0
						Biomass					
1990	2.9	1.7	0.1	0.2	1.3	7.9	0.5	7.6	3.4	11.1	36.7
2050	3.5	2.3	0.5	0.4	5.0	10.0	1.1	12.4	7.0	15.0	57.1
2100	3.9	2.8	0.5	0.4	10.0	13.0	1.1	12.4	7.0	18.7	69.7
						Fuel oil					
1990	7.2	7.0	4.9	0.8	1.3	5.3	3.7	0.2	0.7	0.2	31.2
2050	0.0	0.0	0.0	0.0	0.0	0.0	0.0	0.0	20.8	9.6	30.4
2100	0.0	0.0	0.0	0.0	0.0	0.0	0.0	0.0	0.0	0.1	0.1
						Hydrogen					
1990	0.0	0.0	0.0	0.0	0.0	0.0	0.0	0.0	0.0	0.0	0.0
2050	0.0	0.0	0.0	0.0	0.0	3.2	0.0	0.0	9.1	21.7	33.9
2100	0.0	0.0	0.0	0.0	15.3	0.5	0.0	0.0	0.0	0.0	15.8
						Electricity					
1990	2.1	0.7	0.3	0.0	0.1	0.0	0.0	0.0	0.1	0.1	3.5
2050	2.1	0.7	0.3	0.0	0.1	0.0	0.0	0.0	0.1	0.1	3.5
2100	11.5	21.1	6.3	0.8	0.1	1.5	18.4	15.8	53.1	103.5	232.2
						Total					
1990	30.7	30.1	9.3	2.1	31.8	16.7	8.1	9.4	22.4	15.3	176.0
2050	35.5	33.5	9.6	1.6	23.9	29.7	22.1	12.4	64.0	78.5	309.8
2100	30.0	26.7	6.8	1.2	28.6	37.8	31.1	40.7	74.0	122.3	399.3

Biomass that is fully utilised for thermal energy production makes a great contribution to total consumption of energy carriers for these purposes (approximately 20% in the whole world). It is used in all regions, especially in Latin America, Africa and South and Southeast Asia, where its fraction reaches 40–50%.

Utilisation of biomass is much the same in both Scenarios, i.e. it does not depend on constraints on CO_2 emissions. This is due to the fact that combustion of biomass does not lead to increase in CO_2 concentration in the atmosphere, since at its regeneration the same quantity of carbon is absorbed (assuming complete reproduction of biomass and neglecting decrease in vegetation cover, in particular deforestation of the earth surface).

Table 8.5b. Use of energy resources and energy carriers for heat production, million TJ/year (Scenario 8 — LSM)

Year					Region						
	NA	EU	JK	AZ	SU	LA	ME	AF	CH	SA	World
Natural gas											
1990	18.1	10.5	0.3	0.6	20.3	3.1	3.5	0.1	0.5	1.1	58.0
2050	15.6	16.0	7.1	0.9	18.8	10.8	13.8	11.2	16.9	20.3	131.2
2100	15.2	13.0	3.6	0.6	17.9	6.7	17.2	9.3	23.7	23.9	131.0
Coal											
1990	0.4	10.3	3.8	0.6	8.8	0.4	0.3	1.5	17.8	2.9	46.8
2050	5.5	8.5	0.0	0.2	0.0	0.0	0.0	6.1	17.7	8.1	46.1
2100	5.5	8.5	2.2	0.2	0.0	6.0	0.0	15.7	27.2	28.2	93.4
Biomass											
1990	2.9	1.7	0.1	0.2	1.3	7.9	0.5	7.6	3.4	11.1	36.7
2050	3.0	2.3	0.5	0.4	0.4	10.0	0.9	12.4	7.0	15.0	51.9
2100	3.9	2.8	0.5	0.4	7.9	3.0	1.1	12.4	7.0	18.7	67.6
Fuel oil											
1990	7.2	7.0	4.9	0.8	1.3	5.3	3.7	0.2	0.7	0.2	31.2
2050	3.1	0.0	0.0	0.0	0.0	0.0	0.0	0.0	6.2	15.9	25.2
2100	1.1	0.0	0.0	0.0	0.0	0.0	0.0	0.0	0.0	3.7	4.8
Electricity											
1990	2.1	0.7	0.3	0.0	0.1	0.0	0.0	0.0	0.1	0.1	3.5
2050	2.1	0.7	0.3	0.0	0.1	0.0	0.0	0.0	0.1	0.1	3.5
2100	2.1	0.7	0.3	0.0	0.1	0.0	0.0	0.0	0.1	0.1	3.5
Total											
1990	30.7	30.1	9.3	2.1	31.8	16.7	8.1	9.4	22.4	15.3	176.0
2050	29.4	27.5	7.9	1.5	19.3	20.8	14.7	29.8	47.8	59.3	257.9
2100	27.8	25.0	6.5	1.1	25.9	25.7	18.3	37.5	58.0	74.6	300.3

Fuel oil is used for thermal energy production in small amounts in three regions only – North America, China and South and Southeast Asia, experiencing lack of the other energy resources. On the whole, its use in the 21st century in the world falls as compared to 1990.

Under favourable conditions electricity (see Table 8.5b) must be consumed for heat production on a small scale that will not exceed its use in 1990 (this was prespecified in the runs on GEM-10R as a lower bound on its utilisation for these purposes). Such computation results indicate inefficiency of using electricity for heat supply in ordinary conditions (though in some cases this may prove to be economical). On imposing rigid constraints on CO_2 emissions (see Table 8.5a) electricity turns out to be the key energy carrier for thermal energy production in the end of the century. It is used on an especially large scale in the last two regions (CH and SA). Such scales seem to be unrealistic, like the whole structure of world energy obtained for this Scenario (this was dealt with in Section 7.1).

Under unfavourable conditions (Scenario 5) two new energy carriers — nuclear energy and hydrogen — appear. Nuclear energy, as noted in Section 7.4, is used for heat production in two regions only — Europe and the former USSR, where centralised heat supply is possible from nuclear combined heat and power plants. Clearly this can be realised only if the problem of nuclear reactor safety is successfully solved.

In *mechanical energy* production (Tables 8.6a and 8.6b) regional features manifest themselves to a much lesser extent than for two previous forms of final energy.

Gasoline (generalising all kinds of motor fuel) remains the main fuel for transport over the whole 21st century virtually in all regions, though its share declines by the end of the century. Under favourable conditions (Scenario 8) gasoline consumption grows progressively in developing regions (except for the LA region) in accordance with the growth of their demands for mechanical energy. In the rest of the regions it is partially replaced by other energy carriers in the second half of the century (in Scenario 5 this replacement speeds up).

Natural gas is used on a relatively small scale in all regions starting in 2025 (that is not shown in Tables 8.6a and 8.6b, see Table 7.14 in Section 7.4). It must be noted that in runs on GEM-10R the use of natural gas for mechanical energy production was limited (from above), reflecting actual capabilities for introduction of "gas" technologies. For most regions the optimal solution appeared to be at this boundary, therefore the figures on natural gas for both Scenarios (in both tables) coincided for all regions except for CH and SA in 2050. This indicates economic efficiency of using natural gas in transport in all regions, except for the two mentioned ones that do not possess sufficient gas resources (and are located far from the gas exporting regions). Hence, the potentialities of a wider gas utilisation in transport call for additional studies.

Under favourable conditions the use of methanol (see Table 8.6b) proved to be desirable in North America, Europe, Japan, the former USSR and Latin America, though on a relatively small scale. With more rigid constraints on CO_2 emissions (see Table 8.6a) the mix of regions using methanol somewhat changed, but the total volumes (for the world as a whole) remained small, as far as methanol contains carbon.

Note that for both Scenarios methanol is utilised only for mechanical energy production (see Tables 7.12–7.15 in Section 7.4). And the data on methanol in Table 7.11 (methanol production from natural gas and biomass) present consumption of primary energy for its production. They are higher than in Tables 8.6a and 8.6b, presenting volumes of methanol as such, by the value of losses at its production and further transportation.

Table 8.6a. Consumption of energy resources and energy carriers for mechanical energy production, million TJ/year (Scenario 5 — HRM)

Year	Region										
	NA	EU	JK	AZ	SU	LA	ME	AF	CH	SA	World
Natural gas											
1990	0.0	0.0	0.0	0.0	0.0	0.0	0.0	0.0	0.0	0.0	0.0
2050	3.8	3.1	1.1	0.2	1.5	1.0	0.8	0.2	0.0	0.0	11.7
2100	4.4	4.4	1.6	0.2	2.5	4.2	3.0	2.3	4.1	3.4	30.1
Gasoline											
1990	23.1	14.1	3.6	0.6	9.0	3.1	1.7	1.0	2.9	3.1	62.1
2050	20.0	14.2	0.0	0.5	0.0	9.8	6.4	2.4	27.2	27.7	108.1
2100	14.0	7.2	0.0	0.2	0.0	0.0	0.0	12.0	30.9	32.1	96.4
Methanol											
1990	0.0	0.0	0.0	0.0	0.0	0.0	0.0	0.0	0.0	0.0	0.0
2050	0.0	0.0	2.8	0.0	1.6	0.0	0.0	0.0	0.0	0.0	4.4
2100	0.0	0.0	0.0	0.0	0.0	10.5	7.9	0.0	0.0	21.3	39.6
Hydrogen											
1990	0.0	0.0	0.0	0.0	0.0	0.0	0.0	0.0	0.0	0.0	0.0
2050	2.8	0.0	0.0	0.1	1.2	0.7	0.0	0.0	0.0	0.0	4.8
2100	4.1.	4.0	1.3	0.2	1.7	3.5	2.5	1.5	3.5	2.5	24.9
Electricity											
1990	0.0	0.0	0.0	0.0	0.0	0.0	0.0	0.0	0.0	0.0	0.1
2050	1.4	1.4	0.5	0.1	0.7	0.5	0.3	0.1	0.2	0.2	5.3
2100	2.0	2.0	0.7	0.1	1.2	1.2	1.4	1.1	1.9	1.6	13.2
Total											
1990	23.1	14.1	3.7	0.6	9.0	3.1	1.7	1.0	2.9	3.1	62.2
2050	28.0	18.7	4.4	0.8	5.0	12.0	7.5	2.7	27.4	27.9	134.3
2100	24.5	17.6	3.6	0.7	5.4	19.4	14.8	16.9	40.4	60.9	204.2

The use of electricity for mechanical energy production in Scenario 8 is negligible. Under unfavourable conditions it is coming into use in all regions from the mid-century and, as a rule, in terms of the upper constraint that is set for model runs. This is indicative of its effectiveness in such conditions.

Similarly, under unfavourable conditions (see Table 8.6a) hydrogen should be used in all regions in transport in the second half of the century, though on a small scale.

As concerns chemical energy, the use of energy carriers for its production has no essential regional peculiarities. In the first half of the century basically gasoline (light fractions after oil refining) is used for these purposes in all regions. In the second half gasoline consumption stabilises or somewhat declines (it continues to grow only in the SA region) and it is replaced by natural gas that covers an increasing demand for this energy form.

Table 8.6b. Consumption of energy resources and energy carriers for mechanical energy production, million TJ/year (Scenario 8 — LSM)

Year					Region						
	NA	EU	JK	AZ	SU	LA	ME	AF	CH	SA	World
					Natural gas						
1990	0.0	0.0	0.0	0.0	0.0	0.0	0.0	0.0	0.0	0.0	0.0
2050	3.8	3.1	1.1	0.2	1.5	1.0	0.8	0.2	0.6	0.3	12.6
2100	4.4	4.4	1.6	0.2	2.5	4.2	3.0	2.3	4.1	3.4	30.1
					Gasoline						
1990	23.1	14.1	3.6	0.6	9.0	3.1	1.7	1.0	2.9	3.1	62.1
2050	18.4	7.5	1.0	0.6	3.0	6.4	7.5	15.2	23.9	27.7	111.1
2100	16.8	10.3	0.5	0.5	1.5	1.3	8.9	19.8	29.8	40.8	130.2
					Methanol						
1990	0.0	0.0	0.0	0.0	0.0	0.0	0.0	0.0	0.0	0.0	0.0
2050	2.0	5.5	1.9	0.0	1.6	3.0	0.0	0.0	0.0	0.0	14.0
2100	2.1	0.5	1.4	0.0	1.4	6.8	0.0	0.0	0.0	0.0	12.2
					Electricity						
1990	0.0	0.0	0.0	0.0	0.0	0.0	0.0	0.0	0.0	0.0	0.1
2050	1.1	0.1	0.1	0.0	0.4	0.0	0.0	0.0	0.0	0.0	0.8
2100	0.1	0.1	0.2	0.0	1.2	1.0	0.0	0.0	0.0	0.0	2.6
					Total						
1990	23.1	14.1	3.7	0.6	9.0	3.1	1.7	1.0	2.9	3.1	62.2
2050	24.3	16.2	4.1	0.7	6.5	10.4	8.3	15.4	24.5	28.0	138.5
2100	23.4	15.3	3.7	0.7	6.6	13.3	11.9	22.1	33.9	44.2	175.1

Chapter 9

ANALYSIS OF CONDITIONS AND REQUIREMENTS OF SUSTAINABLE DEVELOPMENT

Requirements for sustainable energy development (see Chapter 2) are diverse and can not be formalised, i.e. can not be reduced to a fixed set of quantitative indices. Besides, they are interrelated. For example, preservation of the environment conceived as having a negligibly low energy impact on nature can not be achieved with current and future indices of technologies and scales of energy production and consumption necessary for economic development, primarily in developing countries. Hence, the case in point should be the reasonable (in some sense) balance between the benefit (positive effect) and damage which depend on the specific energy structure and scales of development.

The benefits and damages may be investigated at different temporal and spatial levels. In terms of sustainable development such an analysis is to include a global level (the whole world) and consider the long-term future (no less than a century).

This chapter addresses this problem as applied to the challenge of greenhouse gas emissions in energy production. Mitigation of emissions to reduce a negative anthropogenic impact on the Earth climate causes additional expenditures (use of more expensive technologies and energy resources). On the one hand, these expenditures should be sufficient to minimise the negative energy impact on nature and the associated damages, and on the other hand, they should not be prohibitively large as to hamper economic development and growth of the living standard of people.

9.1. Energy and climate change

Several recent decades have revealed a man-induced impact on the Earth climate. Combustion of fossil fuel, change in land use and agricultural production increase concentration of greenhouse gases in the atmosphere (causing its warming) and aerosols (causing its cooling). Joint action of greenhouse gases and aerosols will lead in the future to regional and global changes of climate and climate-related characteristics such as temperature, atmospheric precipitation, soil moisture, sea level, etc., as may be supposed from the available scientific information. In many (or in most) cases the climate changes will be unfavourable for mankind.

In 1988 the World Meteorological Organisation (WMO) and the United Nations Environment Programme (UNEP) established jointly the Intergovernmental Panel on Climate Change (IPCC) with the aim to: 1) assess the available information on climate change; 2) estimate the environmental and socio-economic consequences of climate change; 3) generate strategies of response.

The IPCC First Assessment Report was completed in August 1990 and was used as the base for negotiations on the United Nations Framework Convention on

Climate Change (UNFCCC). As stated in Article 2, its objective is to achieve "stabilisation of greenhouse gas concentrations in the atmosphere at a level that would prevent dangerous anthropogenic interference with the climate system. Such a level should be achieved within a time frame sufficient to allow ecosystems to adapt naturally to climate change, to ensure that food production is not threatened and to enable economic development to proceed in a sustainable manner." The Convention was opened for signing at the summit in Rio de Janeiro in 1992.

In December 1995 the IPCC prepared the Second Assessment Report, at present the third report is under way. The publications of the IPCC and its working groups contain a database on the climate change problem, which was formed from available literature and thoroughly examined by experts and governments. The IPCC reports are prepared and reviewed by more than 2000 experts from all over the world. All the member states of WMO have their representatives in the IPCC and its working groups and their governments approve chapters written and edited by experts.

A short survey of problems concerning the climate change (primarily in terms of determination of energy constraints) based on the Second Assessment Report [59] and the IPCC Technical paper [141] which summarise and generalise current scientific data on this challenge is presented below.

9.1.1. Current anthropogenic impact on climate system

Concentrations of the greenhouse gases carbon dioxide (CO_2), methane (CH_4) and nitrous oxide (N_2O) in the atmosphere have considerably increased since pre-industrial times (1750) from 280 to 360 ppmv (parts per million by volume) for CO_2, from 700 to 1720 ppbv (parts per billion by volume) for CH_4, from 275 to 310 ppbv for N_2O. This increase is explained chiefly by human activity: use of fossil fuels, changes in land use and agriculture. In 1990 emissions of carbon dioxide amounted to 6 Gt C (the mass of carbon dioxide is 2.67 times larger), those of methane — 0.5 $GtCH_4$, and nitrous oxide — 0.013 Gt N. For the time span 1860–1994 the total cumulative emissions of carbon dioxide made up 360 Gt C, 240 Gt C of which was due to utilisation of fossil fuel and 120 Gt C due to deforestation and change in land use practice.

The growing concentrations of greenhouse gases cause an additional warming of the atmosphere and the earth surface (on the average). The radiative forcing (W/m^2) which characterises violation of the energy balance of the system "Earth – Atmosphere" is applied as a measure of the impact of GHGs on the climate. They remain in the atmosphere and influence the climate for a long time. The direct radiative forcing of long-lived greenhouse gases (2.45 W/m^2) has been primarily due to increasing concentrations of CO_2 (1.56 W/m^2), CH_4 (0.47 W/m^2), N_2O (0.14 W/m^2) and halocarbons (0.27 W/m^2) since pre-industrial times.

Tropospheric aerosols formed as a result of fuel combustion and from other sources lead to atmosphere cooling. Their radiative (negative) forcing ranges from – 0.25 to – 1.0 W/m^2. As distinct from the long-lived greenhouse gases, the anthropogenic aerosols remain in the atmosphere for a very short time, therefore this radiative forcing is distributed in the regions unevenly and quickly responds to increase or decrease of emissions.

The global average ground temperature has risen since the end of the 19th century by 0.3–0.6°C. Combined with changes in geographical, seasonal and vertical temperature distribution in the atmosphere it gives grounds to relate climate change to human activity. For the last 100 years elevation of the global sea level has reached 10–25 cm which can be explained by the average global temperature rise.

9.1.2. Greenhouse gases: emissions and concentrations

The relation between anthropogenic emissions of the greenhouse gases (first of all carbon dioxide) and their concentrations in the atmosphere has been revealed by mathematical models describing exchange of gases in the atmosphere with several reservoirs (ocean and ecosystems) which adsorb, assimilate, redistribute and release these gases. The anthropogenic carbon emitted into the atmosphere does not decompose, it is rather added to carbon present there and is involved in the processes of its redistribution among reservoirs. These processes are described by typical time intervals from decades (carbon turnover gases in living plants) to a millennium (carbon turnover in the depth of seas and ponds). They exceed many-fold the time during which an individual molecule of CO_2 is present in the atmosphere (it is equal roughly to four years). Because of an essential difference in typical time intervals for different processes of carbon exchange, restoration of the violated equilibrium for CO_2 in the atmosphere can not be described by the common time constant. The results obtained by the climatic models which were applied by the IPCC to establish a relationship between the greenhouse gas emissions and their concentration agree with an accuracy of about 15%.

The modelling results show that if CO_2 emissions remain approximately at the current level, they will lead to practically steady increase of its concentration in the atmosphere at least for two centuries and the resulting concentration will reach a value of about 500 ppmv (i.e. it will almost double as compared to the pre-industrial). Concentration of CO_2 can stabilise at the current level only with immediate reduction of emissions by 50–70% and subsequent further decrease.

Since there is no policy for mitigation of the consequences nor significant technological progress which would reduce emissions and enhance absorption, concentrations of the greenhouse gases and aerosols are supposed to rise during the 21st century. The IPCC has developed six scenarios — IS92 (a–f) — of future emissions of the greenhouse gases and aerosols based on some (greatly varying) assumptions with respect to population growth and the economy, land use, technological changes, accessibility and cost of energy resources for the time span 1990–2100. According to these scenarios, in 2100 carbon dioxide emissions will vary from 6 (the current level) to 36 Gt C per year (from 22 to 132 Gt CO_2 per year) and the low value of the range corresponds to the low growth rates of the population and the economy. Methane emissions will amount to 0.54–1.17 Gt CH_4 per year; those of nitrous oxide — 0.014–0.019 Gt N per year. In all these cases the atmospheric concentrations of greenhouse gases and the total radiative forcing will increase through the considered period.

In addition to the scenarios describing possible (probable under some or other conditions of economic development) greenhouse gas emissions the IPCC also estimated emissions profiles (emissions as a function of time) to stabilise CO_2 concentrations at levels of 450, 550, 650, 750 and 1000 ppmv. This stabilisation will be possible during the time span from 2050 (the concentration is 450 ppmv) to 2350 (the concentration is 1000 ppmv). For the indicated five levels of concentration stabilisation, CO_2 emissions will grow in the first decades of the 21st century against the current level. In the year 2050 they will be equal respectively to 5–8, 8–10, 10–12, 12–13 and 14 Gt C per year, in 2100 — 3, 5–7, 9, 12 and 15 Gt C per year (the different numbers for the same concentration correspond to somewhat distinctive rates of stabilisation to be achieved). In all the variants emissions will continue decreasing to the year 2100.

Any final stabilised concentration is determined primarily by cumulative anthropogenic emissions from the present time to the stabilisation moment, rather than by the way the emissions change during this time interval. It means that at some specified value of the stabilised emission, the larger emissions in the first decades demand their more pronounced decrease at a later time period. The cumulative emissions from 1991 to 2100 which correspond to different levels of concentration stabilisation are presented in Table 9.1 along with the cumulative carbon dioxide emissions in all IPCC IS92 (a–f) scenarios. It shows that without special measures on mitigation of greenhouse gas emissions, none of the probable scenarios of development can stabilise the CO_2 concentration at a level lower than 450–500 ppmv.

Table 9.1. Cumulative emissions of carbon dioxide in 1991–2100, GtC

Variant	Cumulative emissions
Scenario IS92	
c	770
d	980
b	1430
a	1500
f	1830
e	2190
Stabilisation level, ppmv	
450	630–650
550	870–990
650	1030–1190
750	1200–1300
1000	1410

In the variants studied by the IPCC every level of the carbon dioxide emissions is consistent with some definite values of emissions of other greenhouse gases.

However, their influence proves to be of secondary importance and only correct to some extent the result. The overall effect of all greenhouse gases sometimes is taken into account by introducing an equivalent concentration of CO_2 which results in the same radiative forcing. Studies on the sensitivity of calculation results to changes in the emissions of other greenhouse gases show that the corresponding changes in the equivalent concentration of CO_2 make up about 10–15%.

9.1.3. Climate change

Profiles of the carbon dioxide concentrations are used as input for the climatic models to estimate changes in the global average temperature and the sea level. The calculations show that stabilisation at concentrations of 450, 650 and 1000 ppmv will lead to an equilibrium temperature rise as compared to 1990 by about $1^{\circ}C$ (within 0.5 and $1.5^{\circ}C$), $2^{\circ}C$ (within 1.5 and $4^{\circ}C$) and $3.5^{\circ}C$ (within 2 and $7^{\circ}C$), respectively. The value of equilibrium temperature will be reached after the year 2100 (by the year 2300 for the concentration of 1000 ppmv).

In the IPCC IS92a emissions scenario, supposing the "best estimated" value of climate sensitivity, the models predict rise of the global average ground temperature by the year 2100 approximately by $2^{\circ}C$ against 1990. If the "best estimated" value of ice thawing sensitivity to warming is also assumed, the models make a forecast of the sea level elevation by 50 cm to the year 2100.

The average temperature and sea level are only quantitative indicators of future global climate change. Changes at a regional level may be more noticeable and sizeable for nature and man. However, at present the reliability degree of forecasts on the scale "Hemisphere – Continent" which are made on the climatic models "Atmosphere – Ocean" is much higher than that of forecasts on a regional scale. The regional climate change can be predicted, more likely, qualitatively.

9.1.4. Expenses and damages

A large quantity of information has been accumulated on the sensitivity of specific ecosystems to climate change. However, it can not be combined yet into a reliable quantitative estimation of the total damage (of economic and non-economic character) in relation to concentration stabilisation levels or greenhouse gas emission profiles. The studies performed cover, as a rule, the case of a twofold increase of the equivalent CO_2 concentration and the associated rise of equilibrium temperature by $2.5^{\circ}C$ (an average value from the uncertainty interval). The obtained estimates of damages due to warming (which include expenses for their prevention or adaptation to changing conditions) are either qualitative or quantitative pointwise values with a wide scatter.

Warming will bring about changes in the mix and allocation of many ecosystems (forests, pastures, ranges of mountains, deserts, lakes, moistened lands and oceans), as particular species respond to climate change. The biological variety will be reduced, goods and services provided by the ecosystems for society will decrease. Some ecosystems will not be able to reach a new equilibrium for several centuries to come after the climate reaches a new balance.

Climate change will cause intensification of the global hydrological cycle and can have an essential influence on regional water resources. Relatively small changes in temperature and precipitation can give rise to great changes in flows, particularly in arid regions. In deserts the conditions will become, most probably, more extreme, i.e. it will become warmer there, yet not damper. An irreversible process of desertification will be typical of arid areas.

Crop capacity and changes in agricultural productivity will differ essentially by region, resulting in changes in production structure. Productivity will rise in some regions and decrease in others, especially in the tropics and subtropics. The risk of starvation may increase in some regions.

Population of some coastal areas will be more vulnerable to floods and land erosion. With the current level of protection systems the loss of land is estimated at 1% for Egypt, 6% for the Netherlands and 17.5% for Bangladesh.

Climate changes may lead to unfavourable consequences for human health. Increasing mortality of heart diseases due to overheating is an example of direct impact. Indirect impacts will comprise spread of infectious diseases (malaria, yellow fever, encephalitis, etc.) due to expansion of zones for the spread of disease carriers.

The damage caused by climate change may be roughly estimated at several percent of the world GDP (most probably 1.5–2.0%) with the following distribution of this damage between the groups of countries: 1–1.5% for developed and 2–9% for developing countries.

The developing countries will suffer more from climate change because of stronger dependence of their economy on agricultural production and weakness of infrastructure, which does not allow adaptation to unfavourable conditions.

9.1.5. Admissible emission levels

A relationship between greenhouse gas emissions and damages as a consequence of climate change can not be established uniquely based on currently available scientific information. Therefore, optimisation of energy by the "cost-benefit" method, where costs are associated with environmental constraints on its development and benefits with prevented damages to nature and man, is impossible as yet (or at least can not give reliable quantitative results). The energy development options (differing in degree of impact on the environment and additional expenses caused by ecological requirements) can be analysed based on the following qualitative and quantitative formulations:

1. Increase of greenhouse gas emissions because of fossil fuel combustion in the nearest decades is apparently inevitable. During the whole 21st century the greenhouse gas concentration in the atmosphere will grow, leading to global climate change. (Its stabilisation would require an immediate sharp decrease of emissions.)

2. It is necessary to undertake measures (technical, economic and political) to mitigate emissions in comparison with the scenarios without limitation. The priority

measures should be aimed at emission reduction with minimum expenses or even without additional expenses (energy conservation, introduction of new, more economically efficient technologies with lower greenhouse gas emissions).

3. When analysing long-term energy development options, consideration should be given to a set of diverse constraints on greenhouse gas emissions. We consider the following options to be feasible (probable):

a) options which do not require drastic energy restructuring in the coming 2–3 decades;

b) options in which the cumulative carbon dioxide emissions do not exceed 800–1200 Gt C during the time span 1991–2100 (Scenarios 3–8 of the present work);

c) options whose realisation (compared to the options without emission constraints) will demand additional expenses in the amount of several percent (about 2%) of the world GDP.

9.2. Expenses of energy development and GDP

The eight scenarios of the world's energy development which are presented in Chapters 6–8 differ in energy consumption levels, constraints on CO_2 emissions and nuclear energy use. As a result of calculations performed, the economically optimal (under the given conditions) energy structure of the world and its regions and the corresponding expenses of energy system creation and operation were determined (Table 9.2). In the model GEM-10R expenses on energy (for each of the ten regions) are calculated as a sum of costs for all technologies (capital investments and annual operating costs), discounted to the beginning of the given 25-year period.

Analysis of expenses versus CO_2 emission constraints reveals the following regularities:

1. As compared to 1990 expenses will rise considerably by the end of the 21st century: by a factor of 5.1 with a 3.6-fold increase of final energy consumption (Scenario 1) and by a factor of 3.6 with a 2.6-fold increase of final energy consumption (Scenario 8). In the course of time such a rise in expenses will exceed their growth depending on the constraints imposed (difference among the scenarios). Hence, the energy sector will be much more expensive than the current one at any option of its development.

2. With no special measures on CO_2 emission reduction (i.e. if the problem of man-induced effect on the global climate is neglected), nuclear energy is of minor importance in the world energy system: a moratorium on its development leads to a slight growth of expenses and a 7% increase of the cumulative emissions (Scenarios 1 and 2). However, in terms of the current level of knowledge, the emissions in these scenarios seem to be inadmissibly high and should be reduced at least 1.5–2 times and nuclear energy will play an essential role in this reduction.

Depending on the variants of its development (no constraints, moderate constraints or moratorium — Scenarios 3, 5 and 4) the supplementary expenses of rigid constraints on CO_2 emissions differ almost twofold: they rise by 33, 35 and 69% in 2050 and by 42, 66 and 85% in 2100 as compared to Scenario 1 (no constraints).

Table 9.2. Expenses of energy, CO_2 emissions and nuclear energy use (world as a whole)

Year	Scenario							
	1	2	3	4	5	6	7	8
Expenses, trillion US\$/year								
1990	2.1	2.1	2.1	2.1	2.1	2.1	2.1	2.1
2025	3.4	3.4	3.7	4.3	3.8	3.5	3.3	3.3
2050	5.2	5.3	6.9	8.8	7.0	5.6	4.9	4.8
2075	7.8	8.0	11.8	15.3	13.2	9.3	6.4	6.2
2100	10.7	10.9	15.2	19.8	17.8	12.9	8.4	7.6
CO_2 emissions, GtC								
1990	5.9	5.9	5.9	5.9	5.9	5.9	5.9	5.9
2025	12.6	12.7	8.2	8.2	8.2	9.5	9.0	10.1
2050	16.1	18.1	7.6	7.6	7.6	10.9	10.4	12.3
2075	21.9	22.9	7.1	7.1	7.1	10.9	10.9	13.6
2100	24.7	28.4	6.5	6.5	6.5	10.9	10.9	15.0
1991–2100	1734	1864	800	800	800	1072	1042	1242
Nuclear energy, million TJ/year								
1990	23	23	23	23	23	23	23	23
2025	34	23	128	24	44	112	44	43
2050	102	0	541	0	166	243	119	86
2075	60	0	1042	0	212	420	133	105
2100	236	0	1335	0	266	722	146	148

3. Additional expenses rise nonlinearly with enhancement of the rigidity of constraints on emissions: with reduction of the cumulative emissions by 16% (Scenarios 7 and 8), 38% (Scenarios 1 and 6) and 54% (Scenarios 1 and 3) the expenses will rise in 2100 by 10, 21 and 42%, respectively.

4. Decrease in final energy consumption (by 11% in 2050 and 25% in 2100 — Scenarios 7 and 8 in comparison with Scenarios 1–6) results in an essential reduction of expenses (1.3–2.6 times in 2100) and fulfilment of moderate and soft constraints on CO_2 emissions even with reduction of the expenses relative to the variant with no constraints but with high energy consumption (Scenario 1).

The total expenses on energy development and supplementary expenses (caused by environmental constraints) are unequally allocated among the world regions. Most of the expenses including those of solving the global problems, which is a concern of all mankind, have to be borne by developing countries (Figure 9.1).

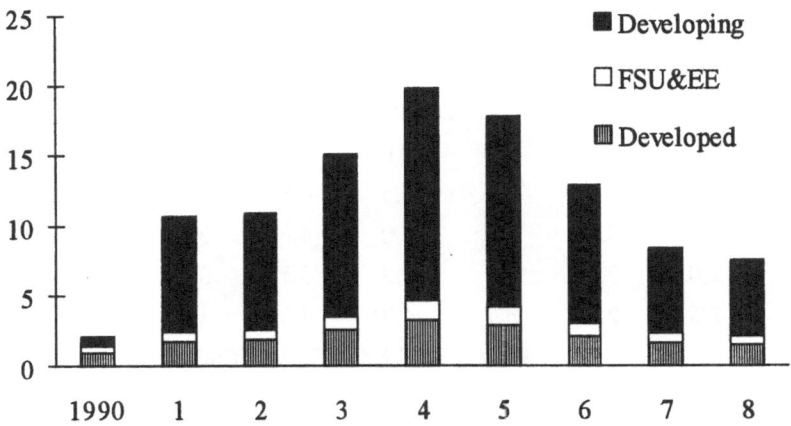

Figure 9.1. Energy expenses of three world regions in 1990 and 2100, trillion US$/year (1–8 are the numbers of scenarios).

In the first approximation the conclusion on inadmissibility of the given level of environmental constraints (in economic terms) can be reached by comparing expenses of energy supply with forecasted GDP values (Table 9.3). Through the 21st century the share of expenses for energy in developed countries, as well as the countries of the former USSR and Eastern Europe, does not exceed the current level and decreases considerably by 2100. For developing countries the rigid constraints on CO_2 in all variants of nuclear energy development (Scenarios 3,4 and 5) may turn out to be unacceptable, since the share of expenses for energy supply in GDP relative to the current level increases, particularly with a moratorium on nuclear energy use (almost by 60% in 2075).

In Scenario 4 supplementary expenses of developing countries (relative to Scenario 1) in the middle and second half of the 21st century make up 4–6% of GDP, which may appear to be more than the supposed economic losses due to climate change. Since a rise in expenses for energy will lead to slower rates of economic growth and impossibility to provide even the required minimum of energy services, developing countries may refuse to impose rigid constraints on emissions.

Scenario 6 (moderate constraints on emissions, no constraints on nuclear energy) seems to be more acceptable among the scenarios with high energy consumption. In this case supplementary expenses relative to Scenario 1 do not exceed 0.6% of GDP in developed countries and 2% in the rest of the world regions. These expenses are less than the forecasted (probable) damages due to climate change, which means that the measures are economically justified. Energy consumption reduction (Scenarios 7 and 8) is the most rational means for reducing greenhouse gas emissions at moderate cost for energy development.

Table 9.3. Expenses for energy of three world regions, % of GDP

Year	Scenario							
	1	2	3	4	5	6	7	8
Developed countries								
1990	7.0	7.0	7.0	7.0	7.0	7.0	7.0	7.0
2025	4.3	4.3	4.7	5.4	4.8	4.4	4.4	4.3
2050	3.6	3.6	4.7	6.0	4.8	3.8	3.8	3.7
2075	3.0	3.1	4.6	5.9	5.1	3.6	3.2	3.1
2100	2.6	2.7	3.7	4.8	4.3	3.2	2.7	2.5
Former USSR and Eastern Europe								
1990	14.0	14.0	14.0	14.0	14.0	14.0	14.0	14.0
2025	9.9	9.9	10.7	12.4	10.9	10.0	10.1	9.8
2050	7.2	7.3	9.5	12.1	9.7	7.7	6.6	6.5
2075	5.7	5.8	8.6	11.2	9.6	6.8	5.7	5.5
2100	5.3	5.4	7.5	9.7	8.7	6.4	5.1	4.6
Developing countries								
1990	7.6	7.6	7.6	7.6	7.6	7.6	7.6	7.6
2025	5.9	5.9	6.4	7.4	6.5	6.0	5.1	5.0
2050	6.1	6.2	8.1	10.3	8.2	6.5	4.9	4.8
2075	6.2	6.3	9.3	12.0	10.3	7.3	4.0	3.9
2100	5.3	5.4	7.5	9.8	8.8	6.4	3.4	3.0

The above trends in the change of expenses depending on the constraints on CO_2 emissions and nuclear energy development are also observed when analysing the expenses per unit of the final energy consumed. Average final energy cost for the world is within the range from US$ 350 (1990) to 850/tce (for Scenario 5 in 2100). At moderate constraints on nuclear energy development (Figure 9.2) rigid constraints on emissions (Scenario 5) cause almost a twofold increase in the energy cost compared to Scenario 1. At low energy consumption, moderate and soft constraints on emissions (Scenarios 7 and 8) practically do not cause any change in the final energy cost. With no constraints on nuclear energy development (Figure 9.3) rigid constraints on emissions (Scenario 3) lead to lesser (compared to Scenario 5) increase in energy cost; moderate constraints on emissions (Scenario 6) affect the energy cost relatively insignificantly up to the mid-21st century.

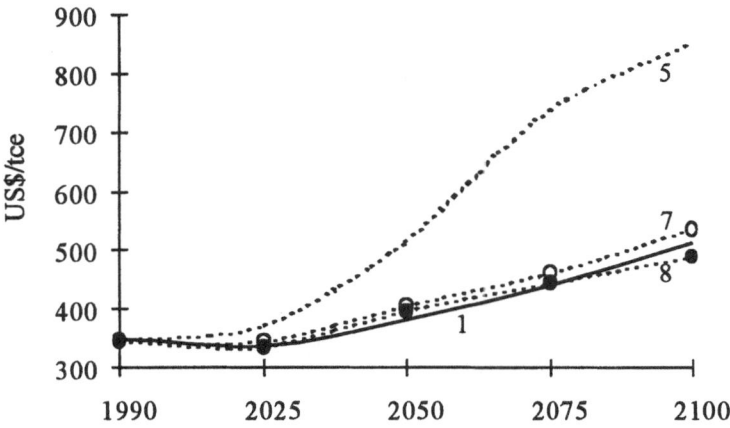

Figure 9.2. Cost per unit of final energy for the world as a whole. Moderate constraints on nuclear energy development, Scenario 1 — without constraints. Figures are Scenarios 1, 5, 7 and 8.

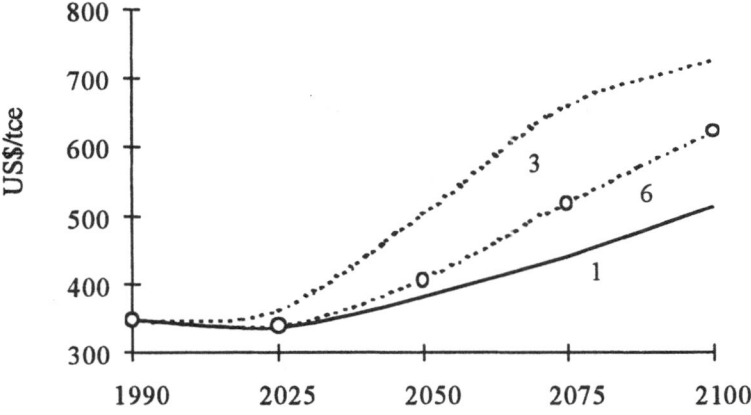

Figure 9.3. Cost per unit of final energy. Without constraints on nuclear energy. Figures are Scenarios 1, 3 and 6.

To analyse the "price" of reducing CO_2 emissions the experts of IPCC often present the data in the form of dependence of supplementary expenses caused by environmental constraints, or economic losses (in percentage of GDP), on the fraction of CO_2 emission reduction relative to the base scenario. Corresponding dependences (the results of calculation on GEM-10R) for the scenarios with high energy consumption are presented in Figure 9.4; Scenario 1 (no constraints) is assumed as the base.

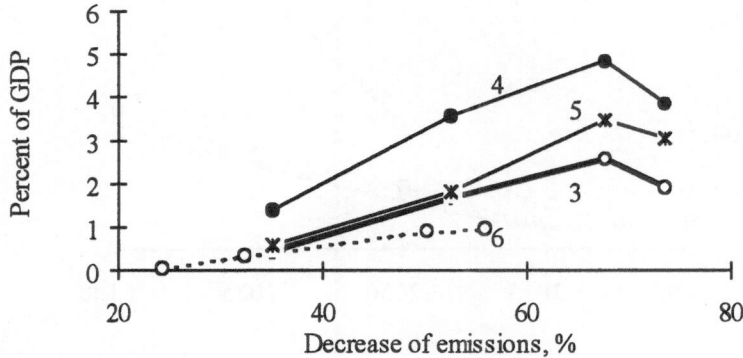

Figure 9.4. Dependence of supplementary expenses of world energy on decrease of CO_2 emissions. Figures are Scenarios 3, 4, 5, 6 relative to Scenario 1.

It is seen that at equal reduction of emissions the expenses increase with growing constraints on nuclear energy and reach maximum at moratorium (4–5% of GDP at emission reduction by 70%). Note, that the results obtained are consistent with the data used by the IPCC experts when preparing the Second Assessment Report [142]. According to this review of numerous works on mathematical modelling of long-term world energy development options (the period considered was from now to 2010, 2020, 2025, 2030, 2050, 2095 and 2100) with the use of different mathematical models (most of them are macroeconomic, applying "top-down" approach), the supplementary expenses lead to GDP losses of 1–2% at decrease of emissions by 40%; at decrease of emissions by 60% most models give losses in the range 1–4% of GDP.

It is possible to calculate supplementary expenses per unit of the prevented carbon emissions into the air (Figure 9.5.) which enables one to tentatively assess the taxes on the fossil fuel consumed by energy sources for creation of an economic mechanism to meet the corresponding constraints. Maximum expenses are reached in Scenario 4 and make up about US$ 500/t C.

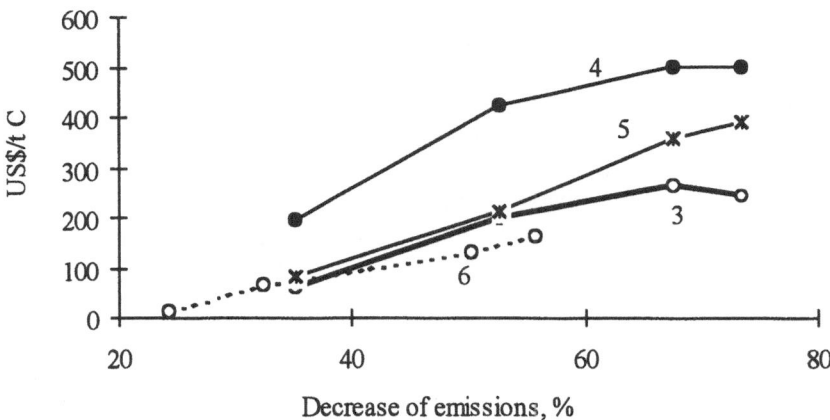

Figure 9.5. Dependence of supplementary expenses on preventing emission of 1t C on decrease of CO_2 emissions. Figures are Scenarios 3, 4, 5, 6 relative to Scenario 1.

9.3. Principles of sharing expenses of CO_2 emission reduction

Transition to specific practical measures on reduction of greenhouse gas emissions requires solving a number of problems associated with organisation of international co-operation in this sphere, first of all, with establishment of national quotas for emissions and rules to regulate the quota exchange and trade. It is well known that this is possible based only on a global approach and co-ordination of actions of all the countries in the world, since efforts to introduce constraints only for individual countries or regions will lead to the so-called effect of "carbon leakage", i.e. to increase of emissions in the other regions [142]. By now, different schemes of quoting greenhouse gas emissions were studied using mathematical modelling, starting with their allocation based on the constant share in the global emission at the date of conclusion of the agreement and ending with the principle of equal rights for emissions for every person [143]. In both these extreme cases international trade of emission quotas is projected in the amount of about US$ 1 trillion by the end of the 21st century, however in the first case the quotas are bought by developing countries and in the second case by developed countries and China.

The negotiations on reduction of greenhouse gas emissions carried out currently in the framework of UNFCCC suppose laying down norms for reduction of emissions relative to 1990 levels for individual countries (first of all for developed). Obviously such an approach (based on the achieved level) in the short-term outlook allows one to more easily obtain the consent of individual countries to assume commitments on emission reduction, the more so that for developed countries, where no considerable energy consumption is planned, such commitments are not economically burdensome. At the same time in the long-term outlook the problem of emission reduction can be solved only by active participation of developing countries that expect the main increase in energy consumption. In doing so it is hard to expect that these countries with per capita CO_2 emissions many times lower than those in developed countries will voluntarily agree to bear expenses associated with reduction of emissions and the more so, to buy quotas for emissions.

Effectiveness of potential strategies of developed countries in the conditions when developing countries will not agree to bear supplementary expenses on reduction of emissions is analysed in the work [144]. Three options are considered:

1) "selfish" scenario — developed countries reduce CO_2 emission to minimise their own economic losses (the total of supplementary expenses on emission reduction and damages due to climate change);

2) "altruistic" scenario — the developed countries reduce their emission to obtain maximum common benefit;

3) "optimal" scenario — all the countries are involved in the measures on reduction of emissions for optimisation of the overall effect, but in doing so developed countries pay to developing (compensate fully for their supplementary expenses on energy).

As could be expected, the "optimal" scenario that offers almost a twofold reduction of global CO_2 emissions (compared to the basic scenario) turned out to be the most effective. Here the net present value of total discounted payments of developed countries to developing makes up about US\$ 0.5 trillion. Let us point out that in the given case energy development options were optimised by the "cost–benefit" method (considering damages due to climate change), therefore the results can hardly be highly accurate (or if at all acceptable for any concrete decisions to be taken).

The mathematical model GEM-10R used to analyse the options of the world energy system development enables one to study allocation of quotas for emissions among the world regions on the assumption of the "ideal" mutually beneficial international co-operation, since the economic energy effectiveness is optimised for the world as a whole. In the model the expenses are allocated between the regions so that their total value for the world as a whole is minimum. Therefore the supplementary expenses for energy of the given region in the general case do not coincide with its real expenses on reduction of global CO_2 emissions. Who exactly has to pay supplementary expenses depends on the chosen criterion. From the view point of sustainable development supposing equality of all the people, the criteria based on the specific (per capita) indices look more reasonable.

In Scenario 1 with no constraints the specific CO_2 emissions for the world as a whole decrease from 1.1 t C per capita in 1990 to 2.5 t C in 2100. There is an essentially uneven distribution of the index by region: in 1990 per capita emissions in developed countries were 9 times and in 2100 were 4 times higher than in developing countries (Figure 9.6). Introduction of moderate constraints (Scenario 6) leads to a 3.3-fold reduction of emissions in developed countries in 2100, a 2-fold in developing countries; introduction of rigid constraints (Scenario 3) leads to a 5.9-fold and 3.7-fold reduction respectively. Despite the greater decrease (relative to the achieved level) of emissions in developed countries their specific emissions in all the options will be higher than in developing countries through the 21st century (Figures 9.7 and 9.8).

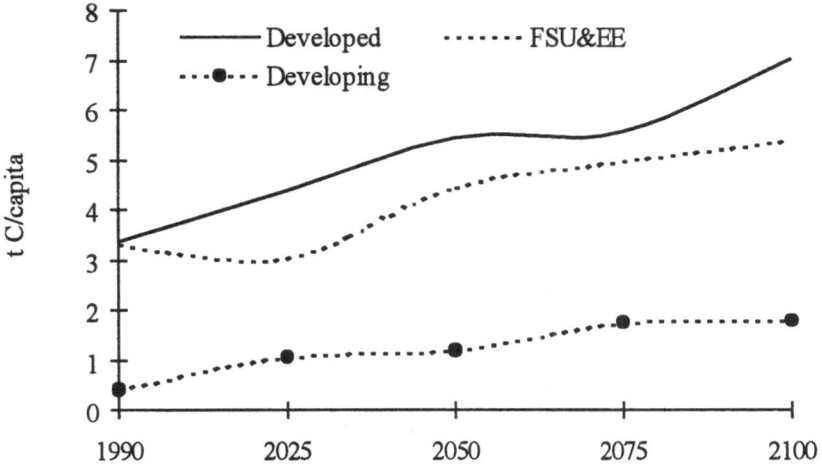

Figure 9.6. Annual per capita CO_2 emissions in three world regions (Scenario 1).

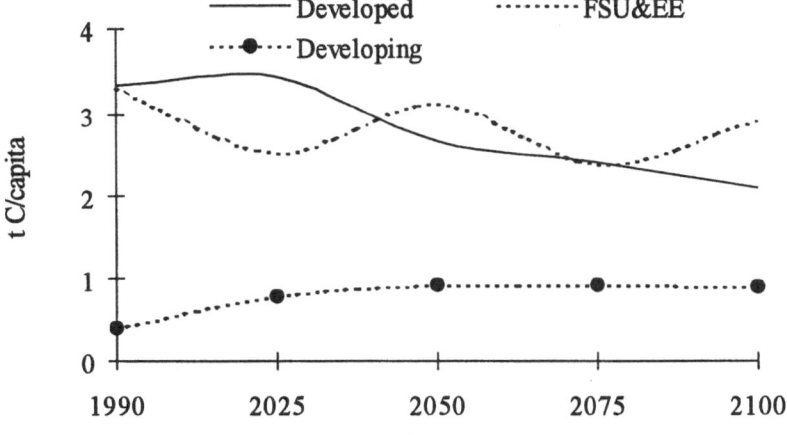

Figure 9.7. Annual per capita CO_2 emissions in three world regions (Scenario 6).

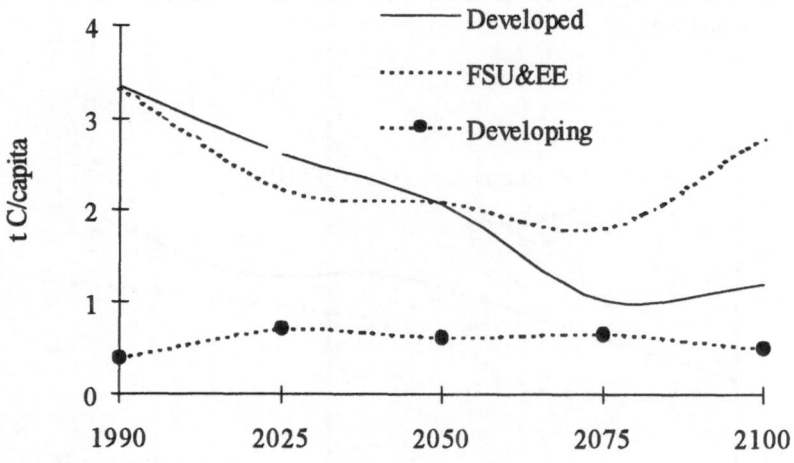

Figure 9.8. Annual per capita CO$_2$ emissions in three world regions (Scenario 3).

In this connection solution of the problem of reducing anthropogenic impact on the climate based on the criterion considering equal rights of all people will make developed countries fully or partially compensate for supplementary expenses of developing countries that reach US\$ 1.5–3.5 trillion/year by 2100 (Figures 9.9 and 9.10).

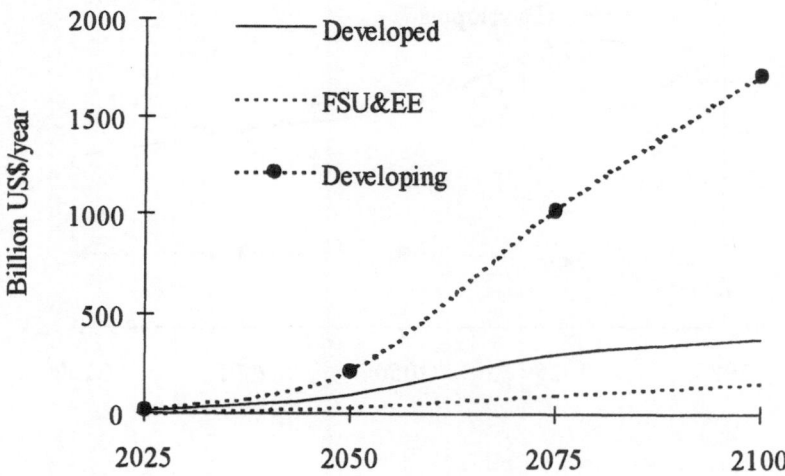

Figure 9.9. Supplementary (relative to Scenario 1) expenses on moderate constraints on CO$_2$ emissions (Scenario 6) for three world regions.

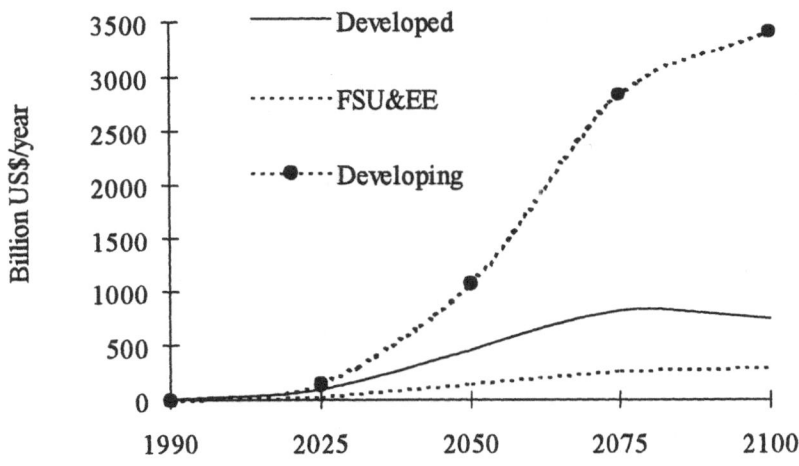

Figure 9.10. Supplementary (relative to Scenario 1) expenses on rigid constraints on CO_2 emissions (Scenario 3) for three world regions.

Determination of specific amount of such a compensation is quite a complicated task: firstly, the global problem of climate change should be analysed on the national basis rather than on the regional (different countries of one and the same region differ greatly in the "price" of emission reduction), which increases the dimensionality a lot; secondly, the price of emission reduction should be determined based on the expenses of energy saving measures (in the model GEM-10R energy consumption is fixed); thirdly, a trade-off criterion of sharing supplementary expenses with regard to interests of all the countries will have to be developed.

Let us consider some possible approaches to determining supplementary expenses on reduction of CO_2 emissions of the regions. To do this it is insufficient to merely calculate the difference of expenses between the basic scenario and the scenario with constraints on emissions for each region, since the calculated expenses relate to the energy facilities located on the territory of the given region, but they service other regions as well, i.e. are a part of the common world energy system.

Let E_i^* and z_i^* (i =1, 2, ... , 10 are the numbers of regions) be respectively CO_2 emissions and expenses obtained from the run on the model GEM-10R. Redistribute these magnitudes (keeping $\Sigma\ E_i^*$ and $\Sigma\ z_i^*$ constant) so that the emissions and expenses correspond to the region's "own" energy, i.e. to the part of it that supplies the region with the predetermined amount of energy to provide the projected economic growth. In the first approximation such redistribution can be performed in proportion to the volumes of the final energy consumed. Thus we obtain new sets of emissions E_i and expenses z_i, here the difference $\Delta z_i = z_i - z_i^{(0)}$ (hereafter the upper index "0" marks the magnitudes related to the basic scenario, the values without index relate to the scenario with global constraints on CO_2 emissions) represents supplementary (related to the emissions reduction) expenses on energy of the i-th region. The total world supplementary expenses are equal to $\Delta z = \Sigma \, \Delta z_i$ and the price of emission reduction is equal to $S = \Delta z / (\Sigma \, E_i^{(0)} - \Sigma \, E_i)$.

However the values Δz_i can not be taken as a "fair" payment of the regions for reduction of emissions either, since the per capita emissions of developing countries are many times less than those of developed countries and they do not have to pay for the problem they have not caused.

Based on the main points of the sustainable development concept, in the given case, on equality of the countries and regions in the right for economic development, the following principles for sharing expenses for prevention of global climate change may be offered.

1. Principle of "ideal justice". Suppose that global reduction is chosen in a way that exceeding the emissions $E_{max} = \Sigma E_i$ will lead to environmental disaster; lesser emissions are safe. Then each inhabitant of the Earth has the right for emissions in the amount of $e = \Sigma E_i / \Sigma P_i$ (P_i is the population number of the i-th region) and eP_i is the quota of the i-th region. Divide all the regions into two groups: A regions that fall beyond the quota allocated for them ($E_i > eP_i$) and B regions that underuse their quota ($E_i < eP_i$). B group consists mainly of the developing countries.

When passing from the basic scenario to a scenario with constraints all the regions (or most of them) reduce their emissions. Nevertheless A regions can not meet the quota and the global constraint E_{max} is accomplished only owing to the fact that the emissions of B regions are less than permitted. This process can be interpreted as the A regions' purchase of quota from B regions, which turns out to be cheaper than to decrease their own emissions. Hence, all the supplementary expenses of the world energy Δz are paid by A regions. Besides, B regions should be additionally paid at the price S for underusing their quota (if they used it fully, for instance, by having increased their energy consumption, A regions would have to additionally reduce their emissions).

Thus, the expenses of B regions (assume that the extra payment is negative expenses) are equal to

$$D_{iB} = - (eP_i - E_i)S. \qquad (9.1)$$

The financial sources (to cover the expenses Δz and payments to B regions) are A regions, whose individual payments are proportional to the excess of their emissions over the set quotas:

$$D_{iA} = (\Delta z - \Sigma_B D_{iB})(E_i - eP_i)/(\Sigma_A E_i - e\Sigma_A P_i). \qquad (9.2)$$

Indices A and B at the sign of the sum indicate the area of summation — either A or B regions.

2. Principle of "real justice". Unlike the previous case, we consider that some B regions in any case will not be able to fully use their quota for emissions eP_i. Assume $E_i^{(0)}$ as a maximum emission level of such a region. Then

$$D_{iB} = - (eP_i - E_i)S \qquad \text{at} \qquad E_i^{(0)} > eP_i$$

and

$$D_{iB} = -(E_i^{(0)} - E_i)S \quad \text{at} \quad E_i^{(0)} < eP_i. \tag{9.3}$$

Expenses of A regions are calculated by formula (9.2).

3. Principle of "compensation". Despite the fact that B regions do not use their quota, they are not paid extra for that. The regions of A group bear only the expenses of technical measures for reduction of global emissions (Δz), i.e. compensate B regions for the supplementary expenses of their energy. In this case

$$D_{iB} = 0, \tag{9.4}$$

and the magnitudes D_{iA} are calculated by the same formula (9.2).

4. Principle of "interest". Up to now we have supposed that the objective of reducing CO_2 emissions is meeting the fixed global constraint. However in fact the purpose is to minimise the total expenses (of preventing climate change and damages caused by it). In this case the optimal level of reduction E_{max} should be chosen so that the total supplementary expenses Δz could be comparable with the expected damage. Suppose that this damage (in cost terms) is proportional to the economy scales of each region, i.e. to the magnitude of its GDP G_i. Then all the regions (of A and B groups) are interested in taking part in the measures on reducing the emissions to minimise their own damage. The share of their participation is approximately equal to the magnitude of the expected damage, i.e.

$$D_i = \Delta z \, G_i / \Sigma \, G_i. \tag{9.5}$$

Table 9.4 presents the results of calculations for four principles of sharing expenses by formulae (9.1)–(9.5) at moderate constraints on CO_2 emissions (Scenario 1 and 6 are compared). It is seen that the regions of AF, SA and ME should become the main receivers of assistance by the principles of "ideal" and "real" justice. China should be rendered insignificant assistance only up to the year 2025.

The principles of "ideal justice" and "real justice" turn out to be absolutely unprofitable for the former USSR and North America. The share of these regions in energy consumption considerably exceeds their share in the world population (1.7–3.5 times) and therefore they top the "fair" quota for CO_2 emissions. On the whole developed countries should spend for measures on emission reduction and on payments to the developing countries 0.1–0.4% of their GDP (US\$ 16–102 billion/year) in 2025 and 1–4.3% of GDP (US\$ 0.7–3.3 trillion/year) in 2100. The least expenses correspond to the principle of "interest", the largest to the principle of "ideal justice" (Figure 9.11).

A considerable share in the expenses of the developed countries is made up by direct financial payments (for exceeding their quotas for emissions) to the developing countries; for the principle of "ideal justice" they increase from US\$ 70 billion in 2025 to 1 trillion yearly in 2100. The first of these magnitudes is close to the financial assistance that was considered when forecasting energy consumption (variant "*L*", Section 3.4) and is easy to realise practically; the second is impossible even theoretically.

Table 9.4. Expenses of the regions for measures on CO_2 emission reduction, % of GDP

Year	NA	EU	JK	AZ	SU	LA	ME	AF	CH	SA	World
				Principle of "ideal justice"							
2025	0.5	0.3	0.2	0.3	1.6	0.0	−0.1	−2.8	−0.1	−0.4	0.1
2050	2.9	1.6	1.3	1.3	6.5	0.3	−0.8	−23.6	0.4	−1.3	0.3
2075	7.3	3.6	3.0	2.6	12.7	1.0	−0.3	−48.5	1.1	−0.7	0.9
2100	7.4	2.8	2.2	1.6	15.4	0.9	−0.4	−18.6	1.1	−0.6	1.0
				Principle of "real justice"							
2025	0.3	0.2	0.1	0.1	0.9	0.0	−0.1	−0.3	−0.1	−0.2	0.1
2050	1.8	1.0	0.8	0.8	4.0	0.2	−0.8	−2.7	0.2	−1.3	0.3
2075	4.5	2.2	1.9	1.6	7.9	0.6	−0.3	−15.8	0.7	−0.7	0.9
2100	7.4	2.8	2.2	1.6	15.4	0.9	−0.4	−18.6	1.1	−0.6	1.0
				Principle of "compensation"							
2025	0.1	0.1	0.1	0.1	0.4	0.0	0.0	0.0	0.0	0.0	0.1
2050	0.9	0.5	0.4	0.4	1.9	0.1	0.0	0.0	0.1	0.0	0.3
2075	2.7	1.3	1.1	1.0	4.7	0.4	0.0	0.0	0.4	0.0	0.9
2100	3.3	1.3	1.0	0.7	6.9	0.4	0.0	0.0	0.5	0.0	1.0
				Principle of "interest"							
2025	0.1	0.1	0.1	0.1	0.1	0.1	0.1	0.1	0.1	0.1	0.1
2050	0.3	0.3	0.3	0.3	0.3	0.3	0.3	0.3	0.3	0.3	0.3
2075	0.9	0.9	0.9	0.9	0.9	0.9	0.9	0.9	0.9	0.9	0.9
2100	1.0	1.0	1.0	1.0	1.0	1.0	1.0	1.0	1.0	1.0	1.0

The results obtained testify that technically possible reduction of greenhouse gas emissions will appear to be hard to implement practically due to contradictions between some countries that differ greatly in levels of economic development. This problem may be a serious obstacle on the way to humanity's transition to sustainable development.

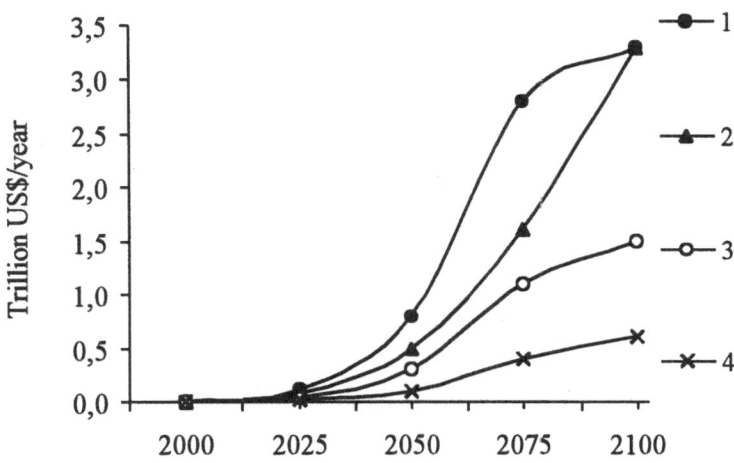

Figure 9.11. Total expenses of developed countries on CO_2 emission reduction in accordance with the principles of: 1 — "ideal justice", 2 — "real justice", 3 — "compensation", 4 — "interest".

9.4. Modelling of relations between energy and the economy

Impacts of additional constraints on world energy development including constraints on CO_2 only as a first approximation can be estimated by change of energy expenses (determined using the model GEM-10R at the fixed energy consumption) and a share of these expenses in GDP. Since the constraints cause a rise in energy cost, a corresponding reorganisation of the economy (reduction in GDP energy intensity) should be expected, energy consumption will decrease and in its turn energy structure and expenses will be changed.

To consider these interactions we will divide the economy of some region into two parts: the energy sector and the rest of the economy (Figure 9.12). The energy sector supplies energy to the economy, whose production output depends on input parameters (energy, labour and capital) and is distributed among the current consumption, investments and energy charge.

Consideration of the interactions between the two sectors allows one to balance energy production and consumption and study the influence of energy cost (and correspondingly constraints on energy development) on energy consumption and GDP.

A similar approach was repeatedly applied earlier. Moreover the combinations of different energy and macroeconomic models [64, 144–147] were used to describe two blocks shown in Figure 9.12. The paper describes the energy sector by the model GEM-10R and the rest of the economy by a macroeconomic model (the modified model MACRO [145]), whose main relations are presented below. The model MACRO was specially developed to study long-term (a 50-year period) energy-economy interrelations. The test calculations show that such a model (with the corresponding combination of parameters) describes very well characteristics for the last years at intervals of at least 15–20 years [147].

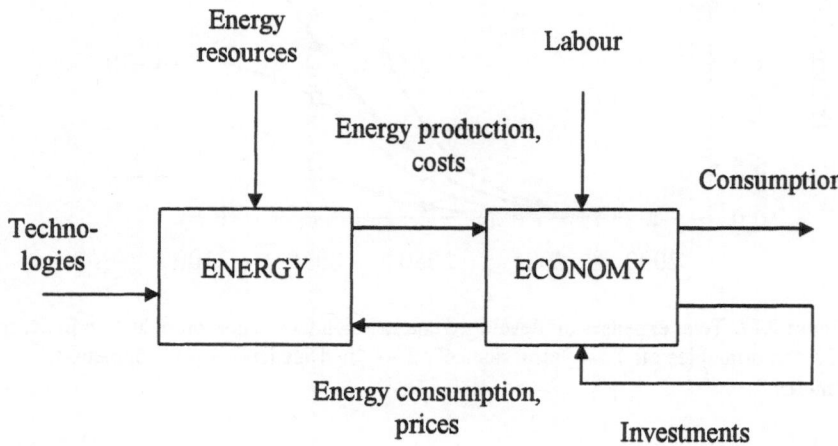

Figure 9.12. Scheme of interactions between the energy sector and the rest of the economy.

All the consumed final energy is divided into two types: electric and non-electric energy (a total of thermal, mechanical and chemical energy). Production output Y depends on labour L, capital K, consumed electric E and non-electric N energy:

$$Y=[a(K^{\alpha}L^{1-\alpha})^{\rho}+b(E^{\beta}N^{1-\beta})^{\rho}]^{1/\rho}, \qquad (9.6)$$

where α and β are elasticity coefficients and constant ρ is connected with the coefficient of substitution σ by dependence $\rho = (\sigma-1)/\sigma$. Magnitudes Y, K, L, E and N are time functions t, moreover $L(t)$ (labour productivity index) is dimensionless (in the basic year $L = 1$). Since the energy share in the economy is not large, as a rule, the magnitude of output Y is close to GDP.

Constants a and b are determined from the initial conditions. Equating derivative $\partial Y/\partial N$ with non-electric energy price p in the basic year we obtain for constant b the equation

$$\partial Y/\partial N = p = Y^{1-\rho}bE^{\beta\rho}(1-\beta)N^{\rho(1-\beta)-1};$$

constant a is found from (9.6) by the known values of all magnitudes in the basic year.

Output Y is determined by energy and non-energy factors (constituents in square brackets of formula (9.6)). Coefficient σ is a quantitative characteristic of their mutual substitution, i.e. substitution of energy with non-energy factor at rise in energy price (energy saving at the cost of additional investments). Actually, the following expression for energy intensity can be obtained from (9.6):

$$N/Y \approx \text{const}(b/p)^{\sigma},$$

i.e. at $\sigma > 0$ with increase in energy price p energy intensity declines. Energy intensity decrease (with the rate of about 1% per year, see Chapter 3) non-related to the price change is modelled by explicit dependence of coefficient b on time.

Output Y is distributed among current consumption C, investments I and expenses on energy:

$$Y = I + C + p_E E + p_N N,$$

where p_E and p_N are prices.

Capital is formed by means of investments and depreciates annually at the rate of $1-\chi$:

$$K(t) = \chi^{\Delta t} K(t-\Delta t) + \varepsilon_1 I(t-\Delta t) + \varepsilon_2 I(t).$$

Here constants ε_1 and ε_2 take into account that investments turn into capital with delay (Δt – time step).

The problem is in finding optimal functions of investments $I(t)$ and energy consumption $E(t)$ and $N(t)$ at which the value

$$F = \Sigma(1+d)^{-t} \log C(t)$$

is maximum (a discounted consumption logarithm; d = discount rate).

The model takes into account that the adjustment to the new energy prices is possible only for newly created energy consumers. For this purpose the output (9.6) is presented in the form of two constituents, one of which is connected with the "old" economy, the other with the "new" one. Such an approach is called the model "clay-wax" (in more detail see [64]). It should be noted that the model is not intended for GDP forecast, since the main factor determining economic growth is the given function $L(t)$ (labour productivity index). With this dependence chosen for some main variant it is possible to study energy impact by analysing deviations of the characteristics from the basic values.

The described model was applied: to additionally test energy consumption forecasts made on the basis of energy-GDP ratio (see Section 3.4); to estimate the impact of changing (variously in different scenarios) energy costs of energy consumption; to specify the impact of environmental constraints on GDP. The region aggregating all developing countries (or, which is the same, some conditional region with average characteristics of developing countries) was considered, since as shown above it is precisely for them that the supplementary expenses make up a considerably large share of GDP and may negatively impact economic growth rates. The calculation was made at a step of $\Delta t = 5$ years for the period from 1990 to 2100 for three variants:

1) energy prices (assumed equal to its production costs) were conditionally supposed to be constant through the considered period;

2) and 3) electric and non-electric energy prices rise proportionally with the ratio of the calculated costs (determined on the model GEM-10R) to the final energy consumed in Scenarios 1 (no constraints) and 4 (rigid constraints on CO_2 emissions and moratorium on nuclear energy use) respectively.

Dependence $L(t)$ was chosen so that the GDP growth in variant 1 approximately corresponded to estimations made in Section 3.4. Elasticity coefficient and coefficient of substitution are assumed equal to the average values from uncertainty intervals [147]: $\alpha=0.3$, $\beta=0.3$, $\sigma=0.4$ (these data, relevant to developed countries, are extended to developing ones). Discount rate is $d = 0.05$ 1/year, capital depreciates at the rate of 4% per year ($\chi=0.96$) which corresponds to the concept accepted in the model GEM-10R that implies full replacement of technologies over a 25-year period.

Optimal investments $I(t)$ obtained from the calculations amount to 18–24% of output $Y(t)$; electric energy expenses $p_E E$ in all variants are within the interval of 1–2% and non-electric energy expenses $p_N N$ decrease from 6% in 1990 to 2.3 and 4% of output $Y(t)$ in 2100 in variants 1, 2 and 3 respectively.

With the constant prices (variant 1) the optimal scales of the consumed energy are in good agreement (in some points they practically coincide) with the magnitudes obtained in Section 3.4 on the basis of energy-GDP ratio (high energy consumption) and used as the initial information for the model GEM-10R (Figure 9.13). Rise in energy prices, primarily due to increase in fossil fuel cost and introduction of more expensive technologies (variant 2) will result in energy consumption decline (by 17–18% in 2100, Figure 9.14). Introduction of environmental constraints and moratorium on nuclear energy use (variant 3) will result in additional decrease in energy consumption (in 2100 by 24%). It is also seen from Figure 9.14 that the variant of low energy consumption accepted for calculations on the model GEM-10R (Scenarios 7 and 8) describes well enough the lower bound of energy consumption variation range due to price changes.

Comparison of variants 2 and 3 (Figure 9.15) allows one to estimate the impact of the most rigid (of the considered) constraints on the main economic characteristic — output Y (or, which is nearly the same, GDP). The GDP losses amount to about 3% which is less than the magnitude of 4–6% (see Section 9.2) determined at constant energy consumption. Decrease in losses is explained by the economy's adaptation to rise in energy prices by introducing energy saving measures.

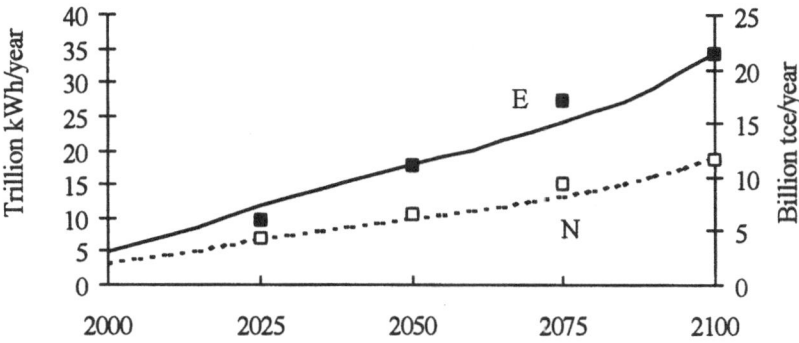

Figure 9.13. Electric E and non-electric N energy consumption in variant 1. Dots are the forecast based on energy-GDP ratio, variant "*H*".

Thus, maximum economic losses (3% of GDP) when introducing rigid environmental constraints that provide cumulative CO_2 emissions of 800 Gt C in the period of 1991–2100 are comparable with or even exceed the projected damage due to global climate change (on average about 2%). These constraints (as shown in Chapter 7) will require a considerable change in the world energy structure (first of all to decrease coal production and use) which will result in a great number of social problems. Thus, the rigid environmental constraints can currently be considered only as one of the variants. The variants with moderate and soft constraints (Scenarios 7 and 8) can be assumed along with rigid constraints as probable scenarios of future energy development (until the corresponding estimates of damages are specified on the basis of new scientific information).

a)

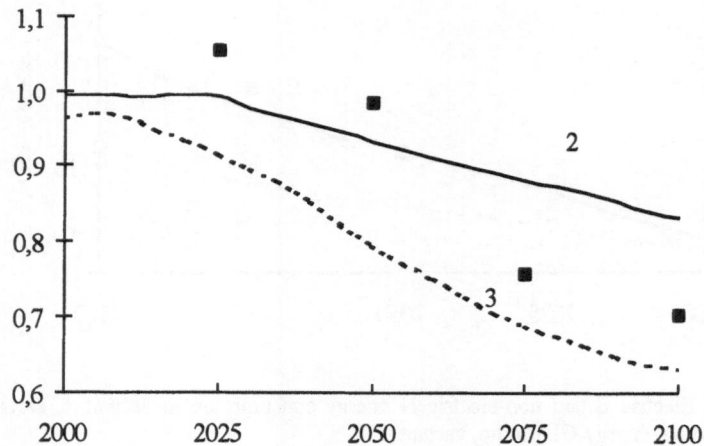

b)

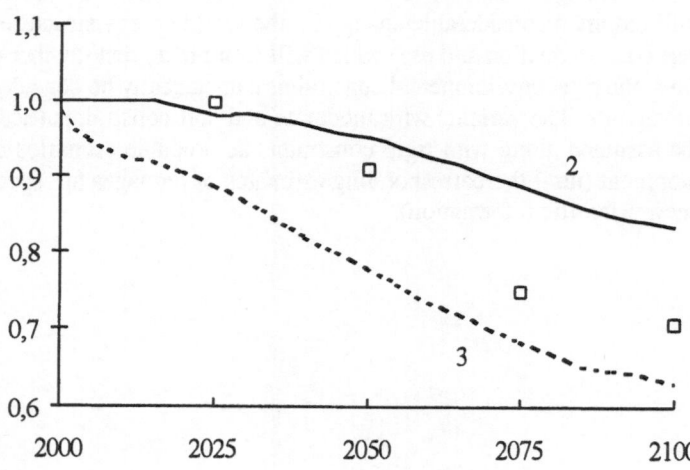

Figure 9.14. Electric (*a*) and non-electric (*b*) energy consumption in variants 2 and 3 in reference to variant 1. Dots are the forecast based on energy-GDP ratio, variant "*L*".

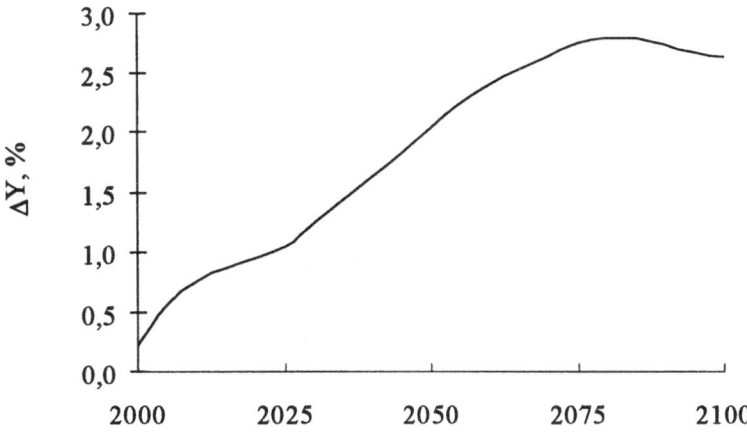

Figure 9.15. Drop in output Y in variant 3 (rigid constraints on emissions, moratorium) in reference to variant 2 (without constraints).

Chapter 10

DIRECTIONS AND PRIORITIES OF
TECHNOLOGICAL PROGRESS IN ENERGY

10.1. Prospects for applying new energy technologies

Analysis of calculations on model GEM-10R for different global scenarios carried out in Chapters 7 and 8 enables one to more thoroughly consider the composition and efficiency of using new energy technologies in the 21st century.

Let us start with the *nuclear energy* technologies. An important point here is the need to develop fast reactors – breeders starting in the mid-21st century (see Table 7.4). In model GEM-10R this technology is taken into account as one using "waste" uranium-238, whose resources can be unlimited, at least in the 21st century (in fact, however, uranium-238 is used in breeders as an additive to plutonium, the basic nuclear fuel produced on a growing scale).

Fast reactors should be applied due to depletion of economically efficient uranium-235 resources at all the considered conditions (scenarios) except for those with a moratorium on nuclear energy development. They are most effective for solving the problem of CO_2 emission reduction. With no constraints on their application (Scenarios 3 and 6) the scales of development turn out to be very large and as we suppose absolutely unrealistic. In Scenarios 5, 7 and 8 with moderate constraints on nuclear energy, development of breeders appears to be more acceptable, particularly in Scenario 8. In this scenario, in 2050 fast reactors are used only in two regions (JK and KT), and in 2100 in all ten regions of the world. The nuclear energy fraction grows significantly (at the expense of breeders) in the region of Japan and Republic of Korea and in all the developing regions (see Figure 8.6–8.10) after 2025 or 2050. Apparently this scenario should be considered as minimum in terms of nuclear energy development.

As was noted in Section 7.1 (see Table 7.6), the "price" of a moratorium on nuclear energy development is very high — energy costs increase by about 40% (or by more than US\$ 4 trillion per year). Therefore *a wide-scaled application of fast reactors can be expected in the second half of the 21st century.* The development of this technology is ongoing — the prototype models and pilot breeders operate in Russia, France, the USA, Japan. It is necessary to accumulate this experience and continue updating to provide opportunities for wide use of breeders in the future.

With stricter constraints on CO_2 emissions it is possible to expect the use of nuclear energy for heat production as well in the regions of Europe and the former USSR (see Tables 7.13 and 8.5a in Scenario 5). First of all it concerns the systems of district heat supply where construction of nuclear cogeneration plants (NCP) may appear to be effective. Construction of nuclear plants for heat production only is much less economical since for several warm months in a year nuclear reactors will stand idle (whereas NCP can produce electric power in the warm period). Hence, in the 21st century the nuclear technology may appear to be needed for combined production of heat and electric power. This technology can not be considered as absolutely new — there are projects of large nuclear cogeneration plants and even a small operating nuclear cogeneration plant in Russia (Bilibinskaya). However, their construction on a larger scale requires higher reliability and safety of nuclear reactors than that of conventional nuclear power plants.

Another way to use nuclear energy for heat supply (industrial) is application of *high-temperature gas cooled reactors* (HTGR). For these reactors Julich Research Center in Germany performed a whole complex of R&D works and a pilot power plant was constructed by the company Brown Boveri in Mannheim. The use of HTGR for electric power production is not economical due to their higher cost compared to conventional nuclear power plants, but they may find an effective application in high-temperature industrial processes, particularly if problems with fossil fuel use indeed arise.

The possibility of using other nuclear energy technologies (thermonuclear reactors, reactors on thorium fuel, etc.) should not be ignored. In the studies described they were not considered since their cost was not supposed to be lower than that of breeders. In the case of successful R&D works at such technologies they may to a greater or a lesser extent replace fast reactors.

The technologies of *renewable energy sources* should be applied on a large scale at rigid constraints on CO_2 emissions: the stricter constraints the larger the scale should be. Let us consider non-traditional technologies of wind, solar, geothermal and space installations and systems that, presently, are not used or used to a lesser extent.

Table 10.1 presents the scales of their application in Scenario 5 (high energy consumption, rigid constraints on CO_2 and moderate on nuclear energy development). Such unfavorable conditions are unlikely but still need to be considered. This scenario addresses quite various types of RES (and their cost categories) therefore it is convenient for analysis of their effectiveness and possible scales of application in different world regions. It should be noted that even in the conditions of low energy consumption and moderate constraints on CO_2 emissions, that are more likely (Scenario 7), RES use can also be great — almost 15 times higher than in 1990 (see Table 7.5 in Section 7.1). The following conclusions can be made analyzing Table 10.1.

In all the calculated periods only the first, the cheapest category *of solar energy* is used. The second and the most expensive third categories (for the average and low level of solar radiation) are not used at all. In this connection, in the 21st century we can assume development of solar energy technologies, which correspond to the first cost category and can be used on a large scale in the regions of North and Latin America, the Middle East and Africa in the second half of the century.

For the first cost category of solar energy as well as for the remaining two, the present studies (on GEM-10R) supposed the photovoltaic method of direct conversion of solar energy into electrical (see Chapters 4 and 5) as applied to the regions with intensive insolation. It is supposed that in the first half of the 21st century photovoltaic cells will become much cheaper. The model envisaged possible energy storage or partial back up of the solar installed capacity by power plants on fossil fuel.

Table 10.1. Use of unconventional RES for electricity production, million TJ/year (Scenario 5 — HSM)

Energy carrier	Region										
	NA	EU	JK	AZ	SU	LA	ME	AF	CH	SA	World
2025											
Helio-1	0.0	0.0	0.0	0.0	0.0	0.0	0.0	0.0	0.0	0.0	0.0
Wind-1	3.0	1.0	0.2	1.0	0.5	0.0	0.1	0.2	0.6	1.2	7.8
Wind -2	27.2	2.5	0.3	1.4	11.5	0.0	0.4	0.0	4.0	3.5	50.8
Wind -3	0.0	3.5	0.0	0.0	10.0	0.0	0.2	0.0	2.0	1.5	17.2
Geo-1	0.0	1.5	0.7	0.0	0.9	0.0	0.3	0.0	0.8	2.5	6.7
Geo-2	0.0	0.0	0.0	0.0	0.0	0.0	0.0	0.0	1.0	5.0	6.0
LPS	0.0	0.0	0.0	0.0	0.0	0.0	0.0	0.0	0.0	0.0	0.0
Total	30.2	8.5	1.2	2.4	22.9	0.0	1.0	0.2	8.4	13.7	88.5
2050											
Helio-1	0.0	0.0	0.0	0.0	0.0	0.0	24.1	0.0	0.3	2.4	26.8
Wind-1	3.0	1.0	0.2	1.0	0.5	1.2	0.1	0.2	0.6	1.2	9.0
Wind -2	27.2	2.5	0.3	0.5	11.5	6.5	0.4	1.2	4.0	3.5	57.6
Wind -3	1.5	3.5	0.3	0.0	10.0	0.0	0.2	0.0	2.0	1.5	19.0
Geo-1	2.5	1.5	0.7	0.4	0.9	0.0	0.3	2.0	0.8	2.5	11.6
Geo-2	0.0	0.5	1.4	0.0	0.0	0.0	0.0	0.0	1.0	7.5	10.4
LPS	0.0	5.9	0.0	0.0	0.0	0.0	0.0	0.0	8.5	10.5	24.9
Total	34.2	14.9	2.9	1.9	22.9	7.7	25.1	3.4	17.2	29.1	159.3
2075											
Helio-1	8.0	0.0	0.0	0.0	0.0	8.0	72.0	17.5	0.0	0.0	105.5
Wind-1	3.0	1.0	0.2	1.0	0.5	1.2	0.1	0.2	0.6	1.2	9.0
Wind -2	27.2	2.5	0.3	4.0	11.5	7.5	0.4	1.2	4.0	3.5	62.1
Wind -3	20.0	3.5	0.3	0.0	10.0	2.5	0.2	2.0	2.0	1.5	42.0
Geo-1	2.5	1.5	0.7	0.1	0.9	2.8	0.3	2.0	0.8	2.5	14.1
Geo-2	0.0	0.5	1.4	0.0	0.0	7.7	0.0	0.0	1.0	7.5	18.1
LPS	0.0	9.2	0.6	0.0	0.0	0.0	0.0	0.0	23.9	33.7	67.4
Total	60.7	18.2	3.5	5.1	22.9	29.7	73.0	22.9	32.3	49.9	318.2
2100											
Helio-1	8.0	0.0	0.0	0.0	0.0	8.0	72.0	40.0	0.0	0.0	127.9
Wind-1	3.0	1.0	0.2	1.0	0.5	1.2	0.1	0.2	0.6	1.2	9.0
Wind -2	27.2	2.5	0.3	3.9	11.5	7.5	0.4	1.2	4.0	3.5	62.0
Wind -3	20.0	3.5	0.3	0.0	10.0	2.5	0.2	2.0	2.0	1.5	42.0
Geo-1	2.5	1.5	0.7	0.0	0.9	2.8	0.3	2.0	0.8	2.5	14.0
Geo-2	0.0	0.0	0.0	0.0	0.0	0.0	0.0	0.0	0.0	0.0	0.0
LPS	0.0	28.8	7.3	0.0	0.0	0.0	0.0	7.5	76.3	145.5	265.4
Total	60.7	37.3	8.8	4.9	22.9	22.0	73.0	52.9	83.7	154.2	520.3

Thus, development of *technologies for direct photovoltaic conversion of solar energy into electrical* should be considered as an important trend in the technological progress in energy.

A steam-power cycle of electric power production with linear parabolic concentrators of solar rays (LUZ technology) which is developed in Israel and the USA and widely applied in California may become another solar energy technology. Currently this technology is noncompetitive, however as fossil fuel becomes more expensive (or its application is constrained) it may appear to be economically efficient. The studies in many countries show that the "tower conception" of solar power plants has proved to have no future.

Wind energy, as is seen from Table 10.1, is used in all the regions up to the third cost category. In 2075 and 2100 each category is used in its full potential. The most expensive wind resources turned out to be noncompetitive. The wind resources are mostly used in North America, the former USSR and Latin America, that possess considerable potential of wind energy.

For all the cost categories the calculations on GEM-10R were performed for one and the same *technology of tower wind mills* with parameters of wind units (number and sizes of blades, speed) that correspond to the wind characteristics. Differences in the cost categories are caused by wind potential (average annual wind velocities and their distribution by duration) and distance from the areas with strong winds to the energy consumption centers in the regions (transmission length). It is necessary to further develop and improve this technology, the more so because the cheapest wind resources are economical in any conditions (see Table 7.5 and 8.4b).

Resources of *geothermal energy* (of the first two categories) are used on the largest scale in 2050 and 2075, particularly in the regions of Latin America and South and Southeastern Asia. In the very end of the century geothermal energy (its second category) is partially replaced by energy from space. On the whole it is quite expensive and competitive only at the most unfavorable conditions (in Scenarios 4, 5, and 7, see Table 7.5).

Geothermal energy resources of the first cost category are determined by the sources of high-temperature subterranean waters (see Chapter 4) and suppose application of *steam power cycles with working media of lower boiling temperatures*. There are such pilot power plants in a number of countries. It is sensible to further improve them. Their application even on a small scale is economically justified in many points of the Earth for any conditions (such details could not be reflected in the global model GEM-10R).

The second category of geothermal energy supposes use of heat from thermal anomalies (that are very rare) with drilling wells and forced circulation of heat-carriers. The expediency of developing such technologies should be determined based on the specific conditions of energy supply in the individual countries. The most expensive third category supposes use of depth heat of hot rocks which can be found everywhere. The calculations show that it is not economical under all the considered conditions (scenarios).

Energy from *Space power systems* (Solar power satellites of the Earth or Lunar power system — see Section 5.5) can be required (turn out to be economically efficient) under the most unfavorable conditions — in Scenarios 4, 5, 7 (see Table 7.5). A very important specific feature of these systems is the need to produce their main elements on the Moon from the materials available there. This requires "making home" on the Moon and constructing necessary production systems there.

In principle it is possible, however implementation of such projects will necessitate considerable time (decades) and large financial expenditures.

The analysis in Section 5.5 showed that two conceptions of Space systems should be considered the most realistic:

1) Solar power satellites (SPS) in geostationary orbit (GSO) [134];

2) Lunar power system (LPS) with additional solar collectors on the back side of the Moon [138].

The studies with calculations on GEM-10R employed two variants of LPS:

— a "simplified" with direct power transmission from the Moon's surface to the Earth (without SHF-reflectors in the orbits around the Earth), that supposes long-term interruptions in power supply;

— a "complete" with SHF-reflectors in the Earth orbits that provide uninterrupted power supply.

The calculations showed that the "simplified" variant of LPS is inefficient in any conditions (scenarios) therefore it can be excluded from consideration.

The results for the "complete" LPS variant are presented in Tables 7.5, 7.12, 8.4a and 10.1. This variant should assume, as is shown in Section 5.5, the LPS with SHF-reflectors in geostationary orbit of the Earth and its economic and energy characteristics will be close to those of the SPS. In this connection two types ("technologies") of Space power systems can be considered promising for further studies and developments:

— *Solar power satellites of the Earth in GSO;*

— *Lunar power system with additional collectors on the back side of the Moon and SHF- reflectors in the Earth's GSO.*

The first stages of R&D works (particularly in developments on the Moon) coincide for both technologies. A justified choice of either technology (when needed) can be made later as the studies proceed and pilot samples are created. Currently it is expedient to consider development of these technologies in frames of the other programs on space development. A real necessity of creating SPS and LPS will be clear after studying the problems of climate warming and change on the planet (including the problems caused by CO_2 emissions).

Biomass as a renewable energy resource does not cause increase in CO_2 concentration in the atmosphere, and is produced in growing amounts almost in all the considered conditions (see Table 7.7). It is used mainly for heat production and as a noncommercial fuel. Besides, under favorable conditions (Scenario 8) it is partially converted into methanol (in the end of the century) and under unfavorable (Scenario 5) into hydrogen (see Table 7.11). It can also be used for electricity production in the beginning of the century (see Table 7.12 for Scenario 5).

Biomass combustion for heating or food cooking is traditional (particularly in the developing countries). There are many similar technologies. They should certainly be developed and improved, first of all increasing their efficiency and automation. However their detailed consideration is impossible here.

The technologies of hydrogen and methanol production from biomass can be considered as new. They will be discussed a bit later along with other technologies of synthetic fuel production.

The above technologies are intended for production of electricity and heat. Table 10.2 presents them ranked (separately for nuclear energy and renewable energy sources) by their economic efficiency and general significance. Such a ranking can not be absolute, therefore the main conditions at which their development and application will be needed are also indicated.

Below there is a list of similar technologies that are considered when calculating on GEM-10R but proved to be inefficient in all the scenarios:

1. Use of fuel oil at base power plants.
2. Electricity production (base and peak) with the use of methanol.
3. Use of hydrogen at base power plants.
4. Geothermal power plants with the use of depth heat of hot rocks.
5. Lunar power system without SHF-reflectors in the Earth's orbits (with interrupted operation mode).

Table 10.2. New top-priority technologies of electricity and heat production

No.	Technology	Conditions of development and application
1.	Nuclear energy technologies	
1.1	Fast reactors (breeders)	In the second half of the 21st century at their due safety
1.2	Combined production of heat and electricity by nuclear reactors	In the systems of centralized heat supply at more rigid constraints on CO_2 emissions
1.3	High-temperature gas cooled reactors	For industrial heat supply at more expensive fossil fuel (or its constrained application)
2.	RES technologies	
2.1	Tower wind installations	Under any conditions
2.2	Direct photovoltaic conversion of solar energy into electric one	A considerable cheapening of photocells, stricter constraints on CO_2 emissions in the regions with higher insolation
2.3	Linear parabolic concentrators of solar rays and steam power cycle	Rise in the cost of or constraints on fossil fuel in the regions with higher insolation
2.4	Steam power cycles with low boiling working media	High temperature geothermal sources, rise in fossil fuel price
2.5	Lunar power systems or Solar power satellites of the Earth	In the second half of the 21st century at stricter constraints on CO_2 emissions in the regions of low insolation or long rain seasons

Let us pass to the technologies of *synthetic fuel production*. Section 7.3 presents the review of their applications for Scenarios 5 and 8 (see Figures 7.3 and Table 7.11). These scenarios cover practically all the range of energy development conditions from the most favorable to the unfavorable. Therefore their analysis allows one to determine the list of different technologies and their efficiency.

Recall that in the calculations on GEM-10R four types of synthetic fuel were distinguished: methanol, substitute natural gas (SNG), hydrogen and motor hydrocarbon fuel produced from coal or methanol. Here consideration was given to the aggregated technologies of their production from different (or with the use of different) primary energy resources, described in Chapter 5 (see also Figure 7.3). Specific technologies (in the framework of the aggregated ones) may be quite various and their choice for different regions will be determined by success of the corresponding R&D works.

The production technologies of synthetic fuel of three kinds are ranked in Table 10.3 (SNG turned out to be inefficient in all the scenarios). This table is made up based on the results for Scenarios 5 and 8 analyzed in Section 7.3 along with the other scenarios that were not considered in detail. Let us make comments on Table 10.3.

Table 10.3. Priority technologies of synthetic fuel production

No.	Technology	Conditions of development or application
1.	Motor fuel production	
1.1	Coal conversion into motor fuel	Under all conditions from 2025 or 2050
2.	Methanol production	
2.1	Methanol production from natural gas	Under all conditions from 2025 or 2050
2.2	Biomass conversion into methanol	In the second half of the century at low energy consumption
2.3	Methanol production from coal	In the first half of the century in the regions poor in oil and natural gas but rich in coal resources (for instance, China)
3.	Hydrogen production	
3.1	With the use of nuclear energy	At rigid and moderate constraints on CO_2 emissions, starting in 2025 or 2050
3.2	Biomass conversion into hydrogen	At rigid and moderate constraints on CO_2 emissions in the second half of the century
3.3	Coal conversion into hydrogen	At rigid constraints on CO_2 emissions in the second half of the century
3.4	Natural gas conversion into hydrogen	Under the same conditions

To produce synthetic motor (hydrocarbon) fuel one more technology was envisaged in the calculations — its production from methanol. However this technology did not work in any of the scenarios. Use of methanol as a motor fuel is more efficient.

Biomass is converted into methanol in Scenarios 8 and 7 (see Table 7.11). At high energy consumption all the biomass is used for production of heat and (or) hydrogen.

Though methanol production from coal was not involved in Scenarios 5 and 8 discussed in Section 7.3, it is used in the others. Therefore this technology is included in Table10.3.

Hydrogen use on a large scale is required at rigid constraints on CO_2 emissions and on much lesser scales at moderate ones (see Tables 7.8, 7.13 and 7.14). At soft constraints on CO_2 emissions (Scenario 8) it is possible to do without it. Therefore the need for developing hydrogen production technologies should depend on the results of solving the problems of warming and change of the planet climate.

Hydrogen production with the use of nuclear energy (including fast reactors) is most efficient (if there is a need). This aggregated technology requires special attention in terms of its application. As Chapter 5 shows different methods are possible here, including thermo-chemical. Creation of special nuclear-energy complexes for production of electric power and hydrogen that may also function as energy storage systems and consumers-regulators seems to be most realistic. Surely the reliability and safety of equipment required, particularly for the use of breeders, should be provided.

The remaining technologies of hydrogen production are known in principle. They are quite various and their development is ongoing. As has been already mentioned, they can be chosen when needed in the first half of the 21st century.

The technologies of fuel processing that turned out to be inefficient under all the considered conditions (scenarios) are: 1) SNG (methane) production from coal and biomass[1]; 2) motor fuel from methanol; 3) hydrogen with the use of electric, solar energy as well as energy from Space power systems; 4) methanol from oil.

The *technologies of final energy consumption* (electrical, thermal, mechanical, chemical) by industry, the residential-commercial sector, agriculture, and transport are even more various; we can say, immense. In GEM-10R they are also represented by the aggregated technologies with averaged technical and economic indices depending on the energy carrier used (see Chapter 5). In Sections 7.4 and 8.2 they are considered at the level of energy carriers on the assumption that each pair "energy carrier — final form of energy" corresponds to an aggregated technology (an electric motor, an internal combustion engine, a boiler unit, an electric heater, etc.). Some electric power consumption technologies are included in "non-substitutable" electricity consumption (for instance, electric lighting). A detailed consideration of such technologies is beyond the scope of our book.

The *concrete fuel production technologies* were not considered either. All of them were aggregated in terms of cost categories of primary fuel resources. Their aggregated representation has been commonly accepted and considered allowable for energy studies at global and regional (a group of countries) levels.

As applied to fuel production an important remark should be made concerning development of *technology for methane production from gas-hydrates* that are found on the bottoms of seas and oceans as well as in the permafrost zones. According to available estimates methane hydrate reserves exceed (in energy equivalent) all the world reserves of the other fossil fuel (coal, oil and natural gas) put together. In the described studies methane hydrates are supposed to belong to the 8-th (the most expensive) category of natural gas resources (unconventional). Therefore their use turned out to be economically inefficient in all the considered scenarios.

However, in the last decade a number of countries (the USA, Japan, Russia, India, etc.) have been conducting rather intensive researches on development of

---

[1] This "global" conclusion does not exclude an efficient production of substitute natural gas in individual countries that have no natural gas but possess cheap coal resources.

technologies for methane production from gas hydrate deposits. There are optimistic estimations on creating such technologies that are based on the first results of the performed experiments (in particular near the shore of Japan and in the Mexican Gulf). If methane production from gas hydrates at the sea bottom becomes economically efficient, this will change significantly the idea of natural gas provision in many countries and regions. This will be a new "breakthrough" in energy comparable with nuclear energy development. This opportunity (scenario) is to be considered in further studies on the global level after obtaining more reliable data on methane hydrate resources and methane production cost.

10.2. Technologically unified multi-product World energy system

The concept of such a system proposed in [6,7] was already mentioned before. In particular it was studied using calculations on the model GEM-10R for the World energy system (WES) type that supposes a large-scale use of Space power systems, particularly the Lunar power system [9,10,12,13]. This section will explain in brief the main idea of such a WES and the results of its studies.

The initial supposition consists in the fact that in the 21st century both centralized and decentralized energy supplies will be developing. No doubt the trend of decentralized RES-based energy supply will become more pronounced, however it is hardly acceptable for cities and large industrial centers. Therefore the centralized energy supply will be developing as well. Large energy systems will continue their formation on the regional and global levels. In fact the existing oil supply system is already of a global character. Regional electric power and gas supply systems have been formed in some parts of the world (in Europe, North America, Asia), and the integration process is ongoing. The hypothesis of a global electric network [148,149] is under consideration. The volumes of interregional transportation of coal and liquefied gas by sea are growing, therefore formation of world systems for electricity-, gas-, coal supply and even nuclear power (at a successful solution of the safety problem) can be expected in the 21st century. The latter will include large international (interregional) facilities for conversion and storage of nuclear fuel and for long-term storage (or burial) of radioactive waste.

These tendencies gave rise to the idea of integrating the indicated specialized energy systems into a multi-product World energy system [6,7]. A similar idea was stated by W.Häfele and W.Sassin before in [57]. The concept below is a further step in this direction and differs from [57] in some aspects. It can become a possible path toward world energy development (for centralized energy supply) in the future.

The multi-product World energy system is expected to have the following features: the WES will represent a single technically and technologically interrelated agglomeration of electricity-, gas-, oil-, coal- and heat supply systems; it will combine the existing world specialized systems or those that may form in the long-distant future; the technological unity of the WES will be its most important property, which will make it necessary to control functioning and development of *the whole* WES. Obviously WES can be created only on the basis of *international* cooperation.

The main elements that will integrate the specialized energy systems and make the WES a technological unit will be large energy centers (with the capacity of tens of GW) intended for combined production of electricity, heat, synthetic fuel, chemical and other products (Figure 10.1). Each center will provide several countries with fuel, energy and raw materials.

The following examples of potential centers can be indicated: gas-chemical complexes for production of electricity, liquid fuel, chemical raw material, etc. that can be created in the Middle East, North Africa, West Siberia; nuclear centers that can produce electricity, synthetic liquid and gaseous fuel and other products from coal, shale, bituminous sands, etc.; energy centers based on the large coal deposits (in Siberia, China, Australia, etc.) for production of electricity and synthetic fuel; large solar-energy centers (in Sahara, Central Asia, etc.) or rectennas (receiving antennas) from Solar power satellites or Lunar power system for production of electric power and synthetic fuel (for instance, hydrogen); similar centers on the basis of large tidal power plants (for instance, Penzhinskaya on the shore of the Okhotsk sea with the capacity of 80 GW).

Such centers will have a powerful transport infrastructure and serve as connecting components between the regional single-product energy systems of different kinds.

Along with the large energy centers (and transport ties) the WES can include: powerful installations (plants, complexes) for conversion of some kinds of fuels and energy into the other (for instance, for hydrogen production by water electrolysis or for electricity production with the use of hydrogen or methane); large-scaled storage systems to   level out daily, seasonal and long-term energy consumption (underground gas storage, big water reservoirs of HPP, hydrogen storage, warehouses for the fuel elements for NPP) that will be used for regulation of power consumption of WES on the whole; for instance underground gas storages can regulate   both gas consumption and electricity production (by gas-fired power plants).

These elements (not shown in Figure 10.1) along with the energy centers create technological unity (interdependence) of the processes in WES and cause the need to coordinate control of the WES development and operation on the whole.

The WES may offer new opportunities to humanity in solving global problems: providing more economic, reliable and stable fuel and energy supply; making easier the solution of environmental problems, caused by the energy sector, including the CO_2 problem; improving energy supply to the developing countries particularly by environmentally clean energy carriers (electric power, natural gas, hydrogen); involving energy resources that are not used currently, first of all, in the developing countries; using an essential effect of combining seasonal (yearly) and daily (weekly) load curves in regions with different altitudes and latitudes; this is particularly important when using RES with non-regular supply; improving territorial location of energy facilities; enhancing economic and business activity, international trade and technological cooperation and improving political climate.

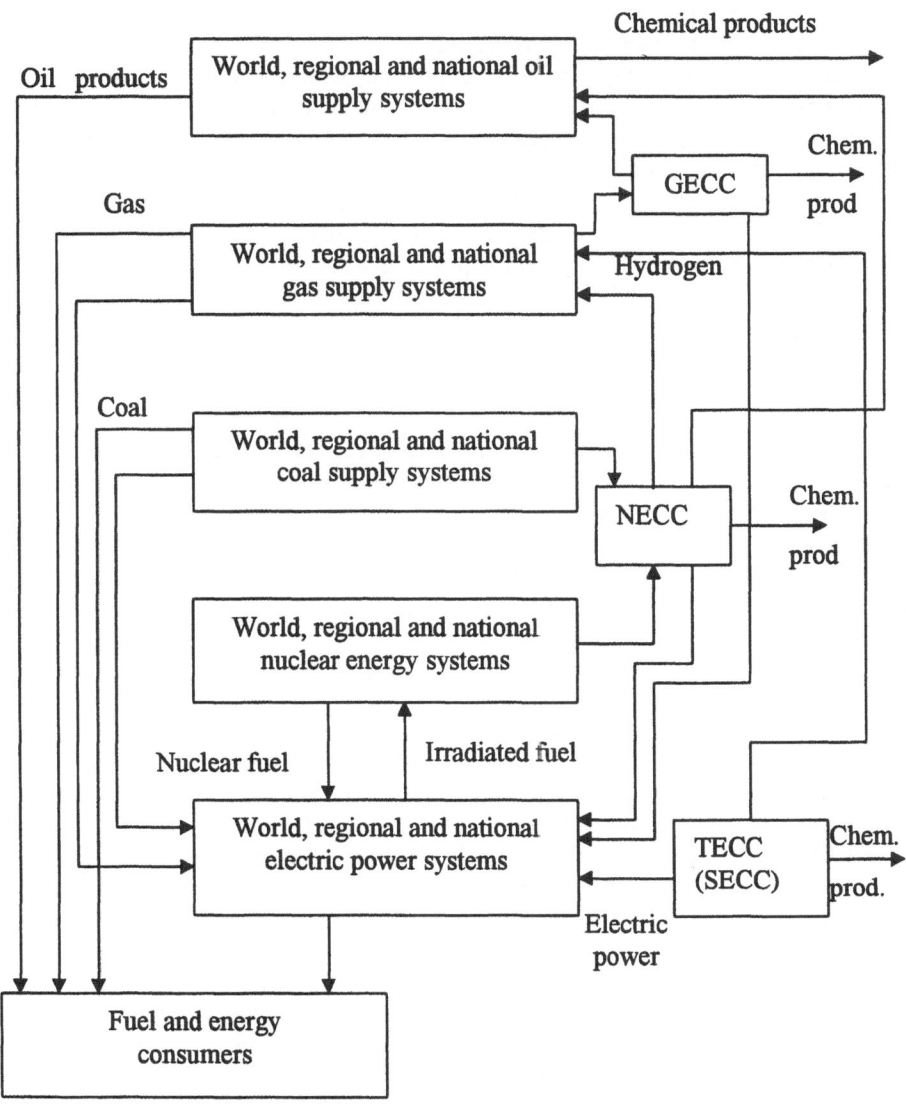

Figure 10.1. The main structure of the World energy system. Energy-chemical complexes: GECC — gas complex, NECC — nuclear complex, SECC — solar complex, TECC — tidal complex.

The WES will apparently be formed gradually (by stages) starting with the current state of national and regional systems of energy. The transition ways will depend on future conditions that seem to be very uncertain now. In the general case within WES we should consider all the forms of energy (oil, gas, coal, electric, solar and nuclear energy), as is shown in Figure 10.1. Depending on the sizes and structure of energy demands, permissible CO_2 emissions and economic indices of different forms of energy (their competition with each other) the scales of development of the specialized systems belonging to the WES will be different (including the state of no development). This will become clear as the future conditions of energy development are clarified.

As studies have shown [9,10,12,13], formation of a technologically unified WES will be quite natural in the case of Lunar power system formation. Energy supply to the Earth from LPS (see Chapter 5.5) will be global indeed in terms of the territory it will cover and mutual connectivity and dependence of energy supply at the different Earth's points. Several rectennas in different parts of the Earth will be supplied in turn from each base of solar collectors (from each antenna) on the Moon. Therefore development and further operation of the WES and LPS should be mutually coordinated.

For instance, rectennas alternately lighted from one and the same antenna, but located in different parts of the WES (on different sides of the Earth) should be of almost (or exactly) equal capacity. At the early stages of LPS creation when choosing the place for and capacity of rectennas it is necessary to take into account further launching of satellites-retransmitters in terms of available sufficient number of lighted antennas on the Moon and impact of reducing (or eliminating) the interruptions in the rectennas operation on the WES.

As LPS develops its operation will become of more and more global character due to interaction of rectennas located in different parts (regions) of the Earth. The operation modes of energy systems in different regions, and hence the WES as a whole, will become more interrelated. With the large capacity of rectennas they can underlie the production of synthetic fuel fed to the liquid and gaseous fuel supply systems. Rectennas will start to play the role of energy centers connecting specialized electric power and fuel supply systems, which was mentioned above.

At the same time the use of space power will become necessary (and even efficient) only under the most severe conditions — at rather rigid constraints on CO_2 emissions and a moratorium on nuclear energy development (see Chapter 7). The probability of introducing such constraints will determine the probability of forming a multi-product World energy system that includes the Lunar power system. By the way, when using Solar power satellites (in GSO) each of them will light only one rectenna that will be similar to an individual power plant of a corresponding capacity. Solar power satellites will not have global interaction of rectennas, which is typical of LPS.

The potentialities of the WES formation based on other kinds of energy centers have not been thoroughly studied yet. Apparently creation of very powerful nuclear, coal and gas energy-chemical centers should be considered almost improbable in terms of safety, reliable energy supply and environment preservation. However foundation of such centers with the use of renewable energy sources (terrestrial solar, tidal and probably some others) may appear expedient, therefore it is planned to continue the study of the WES conception.

CONCLUSION

MAIN RESULTS AND DIRECTIONS OF FURTHER RESEARCH

1. General trends

The results of the studies described here, compared to developments presented by other authors, allow one to determine a number of general trends in world energy development in the century that has just begun.

1. In the 21st century *an essential growth of world energy consumption is inevitable*, first of all in the developing countries. In the industrialised countries, energy consumption may stabilise at current levels or even decrease by the end of the century. According to the low forecast made in the present work, the world consumption of final energy may make up about 350 million TJ/year in 2050 and 450 million TJ/year in 2100 (at the current consumption of about 200 million TJ/year).

2. Humanity *is sufficiently provided with energy resources* for the 21st century, but *an increase in energy price is inevitable*. With 1990 as a base, annual expenditures on world energy will grow 2.5–3 times by the middle of the century and 4–6 times by the end of the century. The average cost of a final energy unit will increase in these periods by 20–30% and 40–80 % respectively (the increase in fuel and energy prices could be considerably more).

3. *Introduction of global constraints on CO_2 emissions* (a most important greenhouse gas) *will greatly affect the energy structure* in the regions and the world as a whole. Attempts to maintain global emissions at the current level should be considered unreal due to a hard-to-solve contradiction: additional costs to limit CO_2 emissions (about US$ 2 trillion/year in the middle of the century and more than US$ 5 trillion/year at the end of the century) will have to be borne mainly by the developing countries which, however, are "not guilty" of creating the problem and do not have the necessary funds; at the same time the developed countries will hardly wish or be able to cover such expenses. Limiting global CO_2 emissions to 40–45 Gt/year in the second half of the century, i.e. to the level of about twice as high as in 1990, can be considered realistic (in terms of satisfactory energy structures in the world regions and costs of its development). In so doing howerever, the problem still exists of sharing the quotas and additional expenses for limiting emissions among the countries and regions.

4. *Nuclear energy development* is the most efficient means for decreasing CO_2 emissions. In the scenarios with rigid or moderate constraints on CO_2 emissions and no constraints on nuclear energy (Scenarios 3 and 6), its optimal development scales appeared to be extremely high. Another index of its efficiency was the "price" of a nuclear moratorium, that, at rigid constraints on CO_2 emissions, causes a 40% increase in world energy costs (more than US$ 4 trillion/year by the end of the 21st century). In this connection our research considers scenarios with "moderate" constraints on nuclear energy development to search for realistic alternatives.

5. Differentiation of *nonrenewable energy resources* into several (eight) cost categories revealed the dynamics of depletion of cheap oil, natural gas, coal and uranium resources in the 21st century and made it possible to forecast fossil fuel prices for a superlong-term period (see Table 7.10). Simultaneously the following points were determined:

— peak world oil production will be reached in the middle of the century; the most expensive (through the sixth to eighth) oil categories will be noncompetitive in the 21st century; production of synthetic motor fuel from coal will be more efficient (starting in the middle of the century);

— world natural gas production in most scenarios increases by the end of the century; two of the most expensive categories representing nontraditional methane and methane hydrates are not used; natural gas use is expedient, first of all, in production of thermal power, peak electric power and synthetic fuel as well as in chemical industry and transport;

— coal resources are depleted to a much lesser extent and increase in coal prices (world prices) will be much less than in oil and natural gas prices; coal consumption particularly depends on the magnitude of global constraints on CO_2 emissions;

— reserves of all eight categories of natural uranium are fully used and fast reactors (breeders) should be used on a larger scale in the second half of the century.

6. Among renewable energy sources the traditional hydropower and biomass as well as the cheapest wind resources are used under all conditions (scenarios). At stricter constraints on CO_2 emissions and on nuclear energy development, the expensive non-traditional renewable energy sources should be used, and the stricter the constraints the larger scale of their use should be.

2. Conditions for transition to sustainable development

The studies described should be considered as the first stage in understanding the energy requirements for sustainable development, in elaborating methodology and in conducting significant studies. The results achieved can be characterised in the following way.

1. The analysis of dynamics of the main economic indices (per capita GDP, per capita energy consumption, etc.) showed that in the 20th century the lag of most developing countries behind the USA (in time of reaching the same values of per capita GDP) has not been reduced and stood at from 50 to 80 years both at the beginning and at the end of the century. If existing economic conditions and interrelations are maintained, such a lag will remain in the 21st century as well (and for the African region will increase). Therefore a necessary condition for transition to sustainable development is *assistance* (financial and technical) *of the developed to the most underdeveloped countries.* To achieve real results such assistance should be rendered in the immediately following decades. Our book presents forecasts of GDP and energy consumption of the world regions that make a case for rendering such assistance.

2. Another condition for transition to sustainable development is providing minimum energy requirement (as consumption increases) for developing countries

and sufficient requirements (as consumption decreases) for the developed ones. Such minimum energy consumption (per capita and in toto) will certainly differ by world regions and will change over time. The condition sets, on the one hand, the hard methodical task of substantiating specific values of minimum or sufficient energy consumption requirements for different regions and time periods and, on the other hand, the problem of practical provision of energy for such consumption. The present monograph offers some approaches and makes the first (preliminary) estimations of energy consumption and the minimum required or sufficient for sustainable development.

3. *The problem of estimating the impact of CO_2 emissions on the planet climate and possibilities to decrease or limit these emissions* remains rather complicated and unclear. In this connection the present work

— determines additional expenses for world energy and its regions for different levels of constraints on global emissions (rigid, moderate, soft);

— compares these costs with possible damage due to climate change (in percentage of world GDP); reveals a tentative equality (comparability) of costs and damages, that enables one to make a conclusion (which should be considered preliminary) on admissibility (or even necessity) of moderating constraints on CO_2 emissions to the values assumed in the corresponding scenarios;

— considers possible principles of "fare" distribution of quotas on CO_2 emissions and additional expenses for reduction of emissions among the world regions.

4. *Energy development options* projected or obtained from calculations on the models *should be tested for "sustainability"*, i.e. conformity with conditions and requirements of sustainable development of humanity. The criteria (indices) and methodology for such a test can not be considered sufficiently understood and developed as yet. A definite step in this direction was made in the present work, in particular, we determined the need:

— to carry out studies at global and regional (group of countries) levels;

— to scrutinise interrelations of energy and the economy (dynamics of GDP, its energy intensity, fraction of GDP consumed for energy development, etc.);

— to analyse energy development options in terms of their practical realisability;

— to perform multi-variant and iterative calculations on different models with thorough and sequential analysis of results to gradually improve the methodology and judge the sustainability of specific options.

Based on these points the energy development options obtained for different scenarios were analysed and two of them (for Scenarios 7 and 8) were selected as the most rational and best meeting the conditions of sustainable development. The options should be further improved to be considered a possible long-term forecast of energy development. Such a forecast could give a more substantiated orientation and basis for taking decisions that require particularly much lead time.

3. Results of studies in the sphere of technological progress

A main objective of the described studies was determination of efficient directions and priorities of technological progress in energy. The main results in this sphere briefly consist in the following.

1. In the 21st century, at least in its first half, fossil fuel will still be mainly used for production of secondary energy carriers and final forms of energy. Therefore it is necessary to keep on updating conventional and developing new technologies to refine oil and convert natural gas and coal.

2. *Nuclear and renewable kinds of energy* will play an important role in the 21st century. Fast reactors (breeders), tower wind installations and photovoltaic cells for direct conversion of solar energy into electric will take a special place among the technologies that use nuclear and renewable energy. In the present work these and other new technologies of power and heat production are ranged in terms of their efficient application (see Table 10.2).

3. Due to depletion of cheap oil resources it will be necessary (and economical) *to produce synthetic fuel in large scales from* 2025 or 2050. The most efficient technologies, which will have to be used under any conditions, are the technologies for synthetic gasoline production from coal as well as methanol production from natural gas, biomass and coal. A large-scale nuclear-based hydrogen production will be required at rigid and even moderated constraints on CO_2 emissions. The conditions and sequence of applying synthetic energy carriers and technologies for their production (see Table 10.3) were determined by the studies.

4. The Space *energy systems* that may appear to be efficient at simultaneous introduction of constraints on nuclear energy development and CO_2 emissions were extensively studied. The modes of energy receiving and technical constraints on application of different Space system concepts have been analysed. Two concepts, in particular, were considered the most promising for further developments. They suppose the use of geostationary orbit (GSO): Solar power satellites and Lunar power system with additional collectors of photovoltaic cells on the back side of the Moon and microwave reflectors in GSO.

5. The concept of *technologically unified multi-product World Energy System* has been studied preliminarily. Formation of such a system will be absolutely natural in the case of creating Lunar power system, since power receiving from the system is really of a global character in terms of the territory covered on the Earth and interrelated modes in different receiving points. Potentialities of the multi-product World Energy System formation on the basis of large nuclear, land-based solar and other energy centres necessitate additional studies.

4. Problems and tasks of further studies

The studies of long-term prospects for energy development in the world, its regions or individual countries will be undoubtedly continued. Each cycle of such studies including the one described in this book leads to better understanding the known problems and at the same time reveals additional tasks or unclear problems.

Particularly difficult problems arise in connection with the necessity of mankind's transition to sustainable development, but there are also other problems both new and already known.

1. *Assistance of developed countries to the developing* is the most important and hard problem, in the authors' opinion. Without such assistance, sustainable development is impossible in general and in the field of energy development in particular. Determination of necessary and possible sizes of such assistance, its distribution among individual countries (both rendering the assistance and getting assistance), creation of mechanisms for its practical realisation, etc. are rather complicated and urgent problems. It is very important for this assistance to be rendered as early as possible, on the one hand, to more quickly bring the living standard of the developing countries closer to that of the developed ones and, on the other hand, for the assistance to make up a considerable fraction in the rapidly increasing total GDP of the developing countries.

2. *Equitable distribution of quotas on CO_2 emissions and costs to decrease emissions among the developed and developing regions and countries* is another complicated problem associated with sustainable development. It is obvious that this problem can and should be connected with the above problem of assistance. However the concrete principles, criteria and mechanisms for allocation of quotas, costs and assistance require comprehension, coordination and approval at the level of international organisations and governments of the countries concerned.

3. Mention should be made of the necessity to continue studies on the general problem of the global carbon cycle and on the effect of anthropogenic CO_2 emissions on the planet climate. Groundless constraints on these emissions would cause great economic damage to mankind. Therefore more comprehensive evidence of the catastrophic character of their effect is required.

4. The problems of *estimating the minimum energy consumption required or sufficient for sustainable development* call for further study that may involve the thermodynamic limits method. An aspect of sustainable development supposing rational utilisation of natural resources supposes minimum (in one sense or another) energy consumption. Comprehension of the criteria, indices and methods for determination and provision of minimum energy consumption is quite a complicated and urgent problem.

5. A more general, non-technical, socio-economic problem consists in *determining the parameters and indices to characterise sustainable development* and, which is most important, *their required quantitative values* for the developed and developing countries in different decades of the 21st century. This is needed to analyse the concrete options of the economy and energy development of countries and regions in terms of their conformity with the principles and criteria of the sustainable development.

6. Among the problems which, for various reasons, proved to be insufficiently studied in the above researches and require further studies, mention should be made of:

— comprehension of rational strategies on consumption of non-renewable energy resources, primarily oil and natural gas, by regions (and countries) — fuel exporters;

— introduction of constraints on energy carrier export-import between the regions based on financial-economic and socio-political conditions; absence of such constraints in the calculations performed led to insufficient use of GEM-10R that could choose the export-import scales within such constraints providing more realistic energy structures of the regions;

— comparison of transport efficiency of crude oil or oil refining products with regard to safety factors and constraints on environmental pollution;

— more thorough analysis of effect of regional constraints on SO_2, NO_x and particulates emissions on the structure and costs of the energy in the regions;

— expediency of imposing regional constraints on CO_2 emissions;

— more comprehensive study of possible and admissible options of nuclear energy development in different regions in the 21st century to impose compromise constraints (in addition to the moderate constraints accepted in the given work for Scenarios 5, 7 and 8).

7. A number of additional studies on the problems associated with technological progress in the energy sector are required:

— inclusion of technologies for combined heat and electricity production at conventional and nuclear power plants in the model GEM-10R and estimation of their efficiency;

— more detailed study of technologies and possible scales of using natural gas in transport;

— study of economic indices and possible scales of applying technologies for natural gas production from methane hydrates;

— continuation of studies on the concept of technologically unified multi-product WES.

REFERENCES

1. **Meadows D. H.** Limits to growth. – New York: Universe Books, 1972.

2. **Häfele W.**, Program Leader. Energy in a finite world: a global systems analysis. – Cambridge, Massachusetts: Ballinger Publ. Comp., 1981.

3. Energy and economy. EMF report. – Stanford, California: Stanford Univ., 1977.

4. **Kononov Yu. D.** Energy and economy: problems of the transition to new energy sources. – Moscow: Nauka, 1981. (in Russian).

5. **Goldemberg J., Johansson T.B., Reddy A.K.N., Williams R.H.** Energy for a sustainable world. – New Delhi: Wiley Eastern Ltd, 1988.

6. **Belyaev L., Papin A., Rudenko Yu., Sinyak Yu.** The multiproduct world energy system: expediency of creation. // Proc. of the 1st Int. Symp. on the World Energy System (St. Petersburg, 4–6 Nov., 1991). – Budapest: Syst. Consulting, Kft, 1992. – P.16–27.

7. **Belyaev L., Rudenko Yu., Sinyak Yu.** The concept of a world energy system // Int. J. Global Energy Issues. – 1994. – Vol. 6, N 3/4/5. – P. 275–279.

8. **Filippov S. P., Kavelin I. Ya.** A mathematical model for the world energy system study // Proc. of the 3rd Int. Symp. on the World Energy System (Uzhgorod, 4–7 Nov., 1993). – Budapest: Syst. Int. Foundation, 1994. – P. 271–280.

9. **Belyaev L. S., Koroteev A.S., Rudenko Yu. N.** Power from Space: expected role and influence on energy system development // Space Power. – 1993. – Vol.12, N 3, 4. – P. 133–142.

10. **Belyaev L. S., Filippov S. P., Kavelin I. Ya.** Evaluation of economic efficiency of the Lunar power system // Proc. of the 3rd Int. Symp. on the World Energy System (Uzhgorod, 4–7 Nov., 1993). – Budapest: Syst. Int. Foundation, 1994. – P. 281–291.

11. **Belyaev L.S., Rudenko Yu. N., Filippov S. P.** Space system concepts: modes of power supply to Earth // Proc. of the Conf. on Alternative Power from Space. – Albuquerque, USA, 1995. – P. 945–949.

12. **Belyaev L. S., Filippov S. P., Marchenko O.V.** Possible role of nuclear energy and power from Space in the 21st century // Proc. of the 1996 World Energy System Conf. – Toronto, Canada, 1996. – P. 143–150.

13. **Belyaev L. S., Filippov S. P., Marchenko O. V.** Possible role of power from Space in the 21st century // Proc. of the SPS'97 Conf. Space Power Systems. Energy and Space for Humanity. – Montreal, Canada, 1997. – P. 35–40.

14. **Belyaev L. S., Filippov S. P., Marchenko O. V., Solomin S. V., Tyrtyshny V. N.** The sustainable energy mixes and the potential role of nuclear power in mitigating CO_2 emissions at regional and global levels. – Irkutsk, Russia: Energy Systems Institute, 1998.

15. **Belyaev L. S., Filippov S. P., Marchenko O. V., Tyrtyshny V. N.** Studies on the potential role of different energy sources in the 21st century // Int. J. of Global Energy Issues (to be published).

16. **Belyaev L. S., Marchenko O.V., Solomin S.V.** et al. Energy of APR in the 21st century on the background of world trends // Proc. of the Int. Conf. Eastern Energy Policy of Russia and Problems of Integration into the Energy Space of the Asia-Pacific Region. – Irkutsk, Russia, 1998. – P. 10–24.

17. **Nakicenovic N., Grübler A., McDonald A.** (eds.). Global energy perspectives. – Cambridge, UK: Cambridge University Press, 1998.

18. **Makarov A.A.** World energy and Euro-Asian energy area. – Moscow: Energoatomizdat, 1998. (in Russian).

19. **Nakicenovic N.** Growth to limits: long waves and the dynamics of technology. – Laxenburg, Austria: International Institute for Applied Systems Analysis, 1984.

20. International energy outlook 2001. – Washington: Energy Information Administration, Office of Integrated Analysis and Forecasting, U.S. Department of Energy, March, 2001.

21. World development indicators 1998. – Washington: World Bank, 1998.

22. BP Statistical Review of World Energy. – London: British Petroleum, 1995 and earlier volumes.

23. **Grübler A., Nakicenovic N., Victor D.G.** Dynamics of energy technologies and global change // Energy Policy. – 1999. – Vol. 27. – P. 247–280.

24. **Fisher J.C., Pry R.H.** A simple substitution model of technological change. Report 70-C-215. – N. Y.: General Electric Company, Research and Development Center, 1970.

25. **Marchetti C., Nakicenovic N.** The dynamics of energy system and the logistic substitution model. RR-79-13. – Laxenburg, Austria: International Institute for Applied Systems Analysis, 1979.

26. **Marchetti C.** Primary energy substitution model: on the interaction between energy and society. WP-75-88. – Laxenburg, Austria: International Institute for Applied Systems Analysis, 1975.

27. **Nakicenovic N.** Patterns of change: technological substitution and long waves in the United States. WP-85-50. – Laxenburg, Austria: International Institute for Applied Systems Analysis, 1985.

28. **Marchetti C.** Nuclear plants and nuclear niches: on the generation of nuclear energy during the last twenty years // Nuclear Sci. and Engin. – 1985. – Vol. 90. – P. 521–526.

29. **Marchetti C.** When will hydrogen come? // Int. J. Hydrogen Energy. – 1985. –Vol. 10, N 4. – P. 215–219.

30. UN MEDS Macroeconomic Data System, MSPA Data Bank of World Development Statistics, MEDS/DTA/1 MSPA-BK.93. – New York: UN, Long-Term Socio-Economic Perspectives Branch, Department of Economic and Social Information & Policy Analysis, 1993.

31. Energy Information Administration. – http://www.eia.doe.gov.

32. US Government Printing Office. – http://www.access.gpo.gov.

33. Global energy perspectives to 2050 and beyond. – London: World Energy Council; International Institute for Applied Systems Analysis, 1995.

34. Penn World Table. – http://www.nber.org/pub/pwt56.

35. World population prospects (the 1992 revision). – New York: UN, 1993.

36. World energy outlook 2000. – Paris: International Energy Agency, 2000.

37. UN statistical reports, 1990 – 1992. New York: UN, 1993.

38. BP statistical review of world energy 1997. – London: British Petroleum. – http://www.bp.com/pubs.html.

39. Energy balances of OECD countries, 1960–1995. – Paris: Organisation for Economic Co-operation and Development, 1996.

40. Energy balances of non-OECD countries, 1960–1995. – Paris: Organisation for Economic Co-operation and Development, 1996.

41. **Edmonds J., Reilly J.M.** Global energy: assessing the future – New York: Oxford University Press, 1985.

42. **Murota Y., Ito K.** Global warming and developing countries: the possibility of a solution by accelerating development // Energy Policy. – 1996. – Vol. 24, N 12. – P. 1061–1077.

43. **Maddison A.** Monitoring the world economy, 1820-1992. – Paris: Organisation for Economic Co-operation and Development, 1995.

44. **Goldemberg J.** A note on the energy intensity of developing countries // Energy Policy. – 1996. – Vol. 24, N 8. – P. 759–761.

45. **Nilsson L.J.** Energy intensity trends in 31 industrial and developing countries 1950-1988 // Energy – The Int. J. – 1993. – Vol. 18, N 4. – P. 309–322.

46. **Doblin C.P.** The growth of energy consumption and prices in the USA, FRG, France, and the UK, 1950–1980. – Laxenburg, Austria: International Institute for Applied Systems Analysis, 1982.

47. Statistical yearbook 1975. – New York: UN, 1976.

48. **Rogner H.-H.** An assessment of world hydrocarbon resources // Ann. Rev. Energy Environ. – 1997. – Vol. 22. – P. 217–262.

49. Oil, gas and coal supply outlook. – Paris: International Energy Agency, 1995.

50. United Nations Conference on Environment and Development, Rio de Janeiro, Brazil, 3-14 June 1992. Earth Summit Agenda 21: The United Nations Programme of Action from Rio. – New York: UN, 1992.

51. **Koptyug V.A., Matrosov V.M., Levashov V.K.** (eds.). A new paradigm of Russia's development in the 21st century. Complex studies on sustainable development problems: ideas and results. – Moscow: Academia, 2000. (in Russian).

52. **Bossel H.** Indicators for sustainable development: theory, method, applications. – Manitoba, Canada: International Institute for Sustainable Development, 1999.

53. **Meadows D.** Indicators and information systems for sustainable development. – Manitoba, Canada: International Institute for Sustainable Development, 1998.

54. Indicators of sustainable development. – New York: UN Department for Policy Co-ordination and Sustainable Development. December, 1994.

55. Working list of indicators of sustainable development. – New York: UN Division for Sustainable Development, 1999.

56. **Marchetti C.** Geoengineering and the energy island. RR-76-1. – Laxenburg, Austria: International Institute for Applied Systems Analysis, 1976.

57. **Häfele W., Sassin W.** The global energy system // Behavioral Sci. –1979. – Vol. 24, N 3. – P. 169–189.

58. **Häfele W., Martinson D., Walbeck M.** The concept of novel horizontally integrated energy systems // Proc. of the X World Congress on Automatic Control. – Munchen: IPAC, 1988. – Vol. 5. – P. 155–161.

59. The IPCC Second Assessment Report on Climate Change. – Geneva, Switzerland: WMO, UNEP, 1995.

60. **Hake J.-F., Kleemann M., Kuckshinrichs W.** et al. Advances in systems analysis: modeling energy-related emissions on a national and global level. – Julich: KFA, 1994.

61. **Belyaev L.S., Filippov S.P., Wagner H.J.** et al. Ways of transition to clean energy use: two methodological approaches. – WP–87–15. – Laxenburg, Austria: International Institute for Applied Systems Analysis, 1987.

62. **Kaganovich B.M., Filippov S.P., Antsiferov E.G.** Energy thechnologies efficiency: Thermodynamics, economics, forecasts. – Novosibirsk, Nauka, 1989. (in Russian).

63. **Desrochers G., Blanchard M., Sud S.** A Monte-Carlo simulation method for the economic assessment of the contribution of wind energy to power systems // IEEE Trans. on Energy Conversion. - 1986. - Vol. 1. - N 4. - P. 50-56.

64. ETA-MACRO: a user's guide. – Stanford, California: Stanford Univ., 1981.

65. **Stepanov V.S.** Analysis of energy efficiency of industrial processes. – Heibelberg: Springer-Verlag, 1992.

66. **Stepanov V.S., Stepanova T.B.** Efficiency of energy utilisation. – Novosibirsk, Nauka, 1994. (in Russian).

67. Uranium: 1995 resources, production and demand. – Paris: Organization for Economic Cooperation and Development, 1996.

68. **Manne A., Richels R.** On stabilizing CO_2 concentration: cost-effective emission reduction strategies. – San Diego: EPRI, 1997.

69. **Demirchian K.S., Demirchian K.K.** A simple model of carbon cycle for upgrading global fossil fuels consumption and carbon emission forecasts validity // Proc. of the Int. Energy Agency Greenhouse Gasses Mitigation Options Conf. (London, 22–25 Aug., 1995). – London: Pergamon Press, 1995. – P. 1265–1270.

70. **Filippov S.P., Tyrtyschnyi V.N.** The GEM-10R software package for the global energy studies // Optimisation methods and their application. – Irkutsk, Russia: Siberian Energy Institute, 1995. – P.306.

71. **Mesarovic M., Pestel E.** Mankind at the turning point. – New York, 1974.

72. **Lapillone B.** MEDEE 2: a model for long-term energy demand evalution. RR-78-17. – Laxenburg, Austria: International Institute for Applied Systems Analysis, 1978.

73. **Stepanov V.S., Stepanov S.V.** Raw material as an energy source // Energy Sources. – 1997. – Vol. 19, N 7. – P. 715–722.

74. **Peterka V.** Macrodynamics of technological change: market penetration by new technologies. RR-77-22. – Laxenburg, Austria: International Institute for Applied Systems Analysis, 1977.

75. Household energy consumption and expenditures 1993. – Washington: Energy Information Administration, Office of Energy Markets and End Use, U.S. Department of Energy, 1995.

76. Annual energy review 1997. – Washington: Energy Information Administration, Office of Energy Markets and End Use, U.S. Department of Energy, 1998.

77. Housing characteristics 1993. – Washington: Energy Information Administration, Office of Energy Markets and End Use, U.S. Department of Energy, 1995.

78. Energy consumption series. Buildings and energy in the 1980's. – Washington: Energy Information Administration, Office of Energy Markets and End Use, U.S. Department of Energy, 1995.

79. **Fickl S.** Energy efficiency in Austria. – http://www.eva.wsr.ac.at /publ/pdf/c33.pdf (April 1999).

80. Commercial buildings characteristics 1992. – Washington: Energy Information Administration, Office of Energy Markets and End Use, U.S. Department of Energy, 1994.

81. Commercial buildings energy consumption and expenditures 1992. – Washington: Energy Information Administration, Office of Energy Markets and End Use, U.S. Department of Energy, 1995.

82. Energy consumption series. Energy end-use intensities in commercial buildings. – Washington: Energy Information Administration, Office of Energy Markets and End Use, U.S. Department of Energy, 1994.

83. Residential transportation at a glance, 1988 – 1994. – Washington: Energy Information Administration, Office of Energy Markets and End Use, U.S. Department of Energy, 1996.

84. **Slavin G.B., Filippov S.P.** Scenarios of external conditions for the world energy system development // Proc. of the 3rd Int. Symp. on the World Energy System (Uzhgorod, 4–7 Nov., 1993). – Budapest: Syst. Int. Foundation, 1994. – P. 293–310.

85. World population prospects 1950 – 2050 (the 1996 revision). – New York: UN, 1996.

86. World population projection to 2150 (long-range projections based on the 1996 revision). – New York: UN, 1998.

87. Total midyear population for the world: 1950 to 2050. – Washington: US Census Bureau, 1998. – http://www.census.gov/ipc/ www/worldpop.html (May 1998).

88. The future population of the world: what can we assume today. – London: Earthscan Publications Ltd, 1996.

89. **Filippov S.P.** World energy resources: cost analysis. – Irkutsk, Russia: Energy Systems Institute, 1994. (in Russian).

90. **Filippov S.P.** Cost analysis of world's hydropower resources and technologies. – Laxenburg, Austria: International Institute for Applied Systems Analysis, 1994.

91. **Astakhov A.** Some estimates of global effective coal resources. Research report. – Laxenburg, Austria: International Institute for Applied Systems Analysis, 1980.

92. **McKelvey V.E.** Mineral resource estimates and public policy // American Scientist. – 1972. – Vol. 60. – P. 32–40.

93. **Rogner H.-H.** An assessment of world hydrocarbon resources. WP-96-56. – Laxenburg, Austria: International Institute for Applied Systems Analysis, 1996.

94. **Modelevski M.S., Gurevich G.S., Khartukov E.M.** et al. Resources of oil and gas and perspectives of their utilisation. – Moscow: Nedra, 1983 (in Russian).

95. **Masters C.D., Root D.H., Attanasi E.D.** World oil and gas resources – future production realities // Ann. Rev. of Energy. – 1990. – Vol.15. – P.23–51.

96. Survey of energy resources: 1992. – London: World Energy Council, 1993.

97. **Bourrelier P.H.** et al. Mobilization of mineral energy resources: a dynamic approach // Proc. of the 14th Congr. World Energy Council. – Montreal, Canada, 1989.

98. BP Statistical Review of World Energy. – London: British Petroleum, 1993, June.

99. **MacDonald G.Y.** The future of methane as an energy resource // Ann. Rev. of Energy. – 1990. – Vol. 15. – P. 53–83.

100. **Grenon M.** On fossil fuel reserves and resources. – Laxenburg, Austria: International Institute for Applied Systems Analysis, 1978.

101. **Cherskiy N., Tsarev V., Nikitin S.** Investigation and prediction of conditions of accumulation of gas resources in gas-hydrate pools // Petroleum Geol. – 1985. – Vol. 21. – P. 65–89.

102. **Sinyak Y., Yamaji K.** (eds.). Energy efficiency and prospects for the USSR and Eastern Europe. – Tokyo, Japan: Central Research Institute of Electric Power Industry (CRIEPI), 1990.

103. Energy statistics yearbook 1991. – New York: UN, 1993.

104. Annual bulletin of electric energy statistics for Europe. – New York: UN, 1993.

105. Statistical yearbook for Asia and the Pacific 1992. – Bangkok, Thailand: UN, 1993.

106. Uranium resources, production and demand. – Paris: International Atomic Energy Agency, 1988.

107. **Slavin G.B.** (ed). Dynamics, structure and resource provision of the world production of primary energy and electricity in 1860–2000. – Irkutsk, Russia: Energy Systems Institute, 1980 (in Russian).

108. **Twidell J.W., Weir A.D.** Renewable energy resources. – London: E. & F. N. Spon, 1986.

109. **Johansson T.B.** et al. (eds.). Renewable energy. Sources for fuel and electricity. – Washington: Island Press, 1993.

110. **Desai A.V.** (ed.). Bioenergy. – New Delhi: Wiley Eastern Ltd, 1990.

111. Energy for tomorrow's world – the realities, the real options and the agenda for achievement. – New York: St. Martin's Press, 1993.

112. **Dessus B., Devin B., Pharabod F.** World potential of renewable energies actually accessible in the nineties and environmental impacts analysis. – Paris: Extraits de la Houille Blanche, 1992.

113. Renewable energy resources: opportunities and constraints 1990 – 2020 // Proc. of the 15th Congr. World Energy Council. – Madrid, 1992.

114. **Piegari H.J.** et al. Ecology and continuation of hydroelectric development in the SIER region // Proc. of the 15th Congr. World Energy Council. – Madrid, 1992.

115. Survey of energy resources. – London: World Energy Council, 1980.

116. **Meyer N.I., Grubb M.J.** Wind power technologies and potentials // Proc. of the Int. Symp. on Environmentally Sound Energy Technologies. – Milan, Italy, 1991. – P.3–39.

117. **Awer P.** et at. Unconventional energy resources // San Diego: EPRI, 1976.

118. **Haraden J.** The status of hot dry rock as an energy sourse // Energy – The Int. J. – 1992. – Vol. 17, N 8. – P. 777–786.

119. **Dessus B., Pharabod F.** Energy development and environment: what about solar energy in a long term perspective? // Proc. of the SPS'91 Conf. Power from Space. – Paris, 1991. – P. 99–108.

120. **Grübler A.** Technology and global change. – Cambridge, UK: Cambridge University Press, 1998.

121. **Ramakumar R., Butler N.G., Rodrigues A.P., Venkata S.S.** Economic aspects of advanced energy technologies // Proc. of the IEEE. – 1993. – Vol. 81. – No. 3. – P. 318–332.

122. **Bakirtzis A.G.** A probabilistic method for the evaluation of the reliability of stand alone wind energy systems // IEEE Trans. on Energy Conversion. – 1992. – Vol. 7. – No. 1. – P. 99–107.

123. **Neij L.** Cost dynamics of wind power // Energy – The Int. J. – 1999. – Vol. 24 – P. 375–389.

124. European wind turbine catalogue. – Copenhagen: Energy Centre Denmark, 1994.

125. **Morthorst P.E., Jensen P.H.** Economic of wind turbines // Wind energy in Denmark: research and technological development. – Copenhagen: Ministry of Energy; Danish Energy Agency, 1990. – P. 54–55.

126. Proceedings of the 9th Photovoltaic Solar Energy Conf. – FRG, 1989.

127. **Benner J.P., Kazmenski L.** Photovoltaics. Gaining greater visibility // IEEE Spectrum. – 1999. – Vol. 36, N 9. – P. 34–42.

128. **Notton G., Muselli M., Poggi P.** Costing a stand alone photovoltaic system // Energy – The Int. J. – 1998. – Vol. 23, N 4. – P. 289–308.

129. Photovoltaic technologies and their future potential. – Berlin: EAB-OPET, 1993.

130. **Wrixon G.T., Rooney A.-M.E., Palz W.** Renewable energy-2000. –Berlin: Springer-Verlag, 1993.

131. **Dickson M.H, Fanelli M**. Geothermal energy worldwide // The world directory of renewable energy suppliers and services 1995. – London: James & James, 1995. – P. 78-83 c.

132. **Anderer J., Häfele W., Barnert H.** et al. The concept of novel horizontally integrated energy systems: the case of zero emissions. Julich: KFA,1984.

133. **Patterson G.H.** The future role of hydrogen fuel in an electrical society. – Toronto: UTUHS, 1979.

134. **Glaser P.** Solar power from satellites // Physics Today. – 1977. – February. – P. 30–38.

135. **Criswell D. R., Waldron R. D.** Lunar system to supply solar electric power to Earth // Proc. of the 25th Intersoc. Energy Conversion Engineering Conf. – Reno, NV, 1990. – Vol.1. – P. 61–70,

136. **Kulcinski G.L.** et al. Fusion energy from the Moon for the 21st century // Proc. of the 2nd Lunar Base Conf. – Houston, TX, 1988.

137. Proceedings of the SPS'91 Conf. Power from Space. – Paris, 1991.

138. **Criswell D.K., Waldron R.D.** Lunar solar power system: options and beaming characteristics // Proc. of the Int. Astronautical Federation Congr. – Paris, 1993.

139. Lunar energy enterprise case study task force. TM-101652. – NASA, 1989.

140. **Criswell D.R., Waldron R.D.** International Lunar base and Lunar-based power system to supply Earth with electric power // Proc. of the 42nd Congr. Int. Astronautical Federation. – Montreal, Canada, 1991.

141. Stabilization of atmospheric greenhouse gases: physical, biological and socio-economic implications. IPCC Technical Paper III. – IPCC, February 1997. – http://www.ipcc.ch/pub/IPCCTP.III(E).pdf.

142. **Richels R., Sturm P.** The costs of CO_2 emission reductions. Some insights from global analyses // Energy Policy. – 1996. – Vol. 24, N. 10/11. – P. 875–887.

143. **Edmonds J., Wise M., Barns D.W.** Carbon coalitions: the cost and effectiveness of energy agreements to alter trajectories of atmospheric dioxide emissions // Energy Policy. – 1995. – Vol. 23, N. 4/5. – P. 309–335.

144. **Peck S.C., Teisberg T.J.** International CO_2 emissions control. An analysis using CETA // Energy Policy. – 1995. – Vol. 23, N. 4/5. – P. 297–308.

145. MARKAL-MACRO: an overview. – USA, Brookhaven Nat. Lab., 1992.

146. Energy, environment and the economy in a CGE model concept. – Amsterdam: Stichting voor Economisch Onderzoek der Univ. van Amsterdam, 1997.

147. **Rogner H.-H.** A long-term macroeconomic equilibrium model for the European Community // IIASA reports. – 1982. – Vol. 5. – P. 255–331.

148. **Rudenko Yu., Yershevich V.** Is it possible and expedient to create a global energy network? // Int. J. Global Energy Issues. – 1991. – Vol. 3, N 3. – P. 159–165.

149. Executive summary // Proc. of the Int. Workshop Global Electric Energy Grid (Winnipeg, Manitoba, Canada, 10–13 July, 1991). – San Diego: GENI, 1991.

ACRONYMS

APR	— Asian-Pacific region
EO	— Earth orbit
EPRI	— Electric Power Research Institute (the USA)
GEM-10R	— Global energy model (for 10 regions)
GDP	— Gross Domestic Product
GSO	— Geostationary orbit
HDD/CDD	— Number of heating/cooling degree—days
HPP	— Hydropower plant
HTGR	— High-temperature gas cooled reactor
IIASA	— International Institute for Applied Systems Analysis
IAEA	— International Atomic Energy Agency
IPCC	— Intergovernmental Panel on Climate Change
LF	— Liquid fuel
LO	— Lunar orbit
LPS	— Lunar power system
m.e.	— Mechanical energy
NCP	— Nuclear cogeneration plant
NE	— Nuclear energy
NFR	— Non-fossil renewables
NPP	— Nuclear power plant
OR	— Oil refinery
OPEC	— Organization of Petroleum Exporting Countries
p.i.	— Power input
PPP	— Purchasing power parity
PVC	— Photovoltaic converter
R&D	— Research and development works
RES	— Renewable energy sources
RNF	— Renewable non-fossils
s.e.	— Secondary energy
SHF	— Super high frequency
SNG	— Substitute natural gas
SPS	— Solar power satellite
SPP	— Solar power plant
tce	— tonn of coal equivalent
t.e.	— thermal energy
TPP	— Thermal power plant
UNEP	— United Nations Environment Programme
WEC	— World Energy Council
WES	— World Energy System
WMO	— World Meteorological Organization
WPP	— Wind power plant

Notations of regions

AZ	— Australia and New Zealand
AF	— Africa
ME	— the Middle East
EU	— Europe
CH	— China
LA	— Latin America
NA	— North America
SU	— former USSR
SA	— South and Southeast Asia
JK	— Japan and Republic of Korea
FSU and EE	— former USSR and East European countries

Notations of scenarios

HNN	— High energy consumption, no constraints on CO_2 emissions, no constraints on nuclear energy development
HNR	— High energy consumption, no constraints on CO_2 emissions, rigid constraints (moratorium) on nuclear energy development
HRN	— High energy consumption, rigid constraints on CO_2 emissions, no constraints on nuclear energy development
HRR	— High energy consumption, rigid constraints on CO_2 emissions, rigid constraints (moratorium) on nuclear energy development
HRM	— High energy consumption, rigid constraints on CO_2 emissions, moderate constraints on nuclear energy development
HMN	— High energy consumption, moderate constraints on CO_2 emissions, no constraints on nuclear energy development
LMM	— Low energy consumption, moderate constraints on CO_2 emissions, moderate constraints on nuclear energy development
LSM	— Low energy consumption, soft constraints on CO_2 emissions, moderate constraints on nuclear energy development

Index

261